Contributions to Environmental Sciences & Innovative Business Technology

Contributions to Environmental Sciences & Innovative Business Technology (CESIBT) is an interdisciplinary series of peer-reviewed books dedicated to addressing emerging research trends relevant to the interplay between Environmental Sciences, Innovation, and Business Technology in their broadest sense. This series constitutes a comprehensive up-to-date interdisciplinary reference that develops integrated concepts for sustainability and discusses the emerging trends and practices that will define the future of these disciplines.

This series publishes the latest developments and research in the various areas of Environmental Sciences, Innovation, and Business Technology, combined with scientific quality and timeliness. It encompasses the theoretical, practical, and methodological aspects of all branches of these scientific disciplines embedded in the fields of Environmental Sciences, Innovation, and Business Technology.

The series also draws on the best research papers from EuroMid Academy of Business and Technology (EMABT) and other international conferences to foster the creation and development of sustainable solutions for local and international organizations worldwide. By including interdisciplinary contributions, this series introduces innovative tools that can best support and shape both the economical and sustainability agenda for the welfare of all countries, through better use of data, a more effective organization, and global, local, and individual work. The series can also present new case studies in real-world settings offering solid examples of recent innovations and business technology with special consideration for resolving environmental issues in different regions of the world.

The series can be beneficial to researchers, instructors, practitioners, consultants, and industrial experts, in addition to governments from around the world. Published in collaboration with EMABT, the Springer CESIBT series will bring together the latest research that addresses key challenges and issues in the domain of Environmental Sciences & Innovative Business Technology for sustainable development.

Bhajneet Kaur · Sudhir Rana ·
Sudhi Sharma
Editors

Integrating Industry 5.0 and 4.0 for Technology and Humanity

Springer

Editors
Bhajneet Kaur
Department of Management Studies, Area Chair (IT and Business Analytics)
Rukmini Devi Institute of Advanced Studies (RDIAS)
New Delhi, Delhi, India

Sudhir Rana
College of Business
Liwa University
Abu Dhabi, United Arab Emirates

Sudhi Sharma
Amity University
Noida, India

ISSN 2731-8303 ISSN 2731-8311 (electronic)
Contributions to Environmental Sciences & Innovative Business Technology
ISBN 978-981-92-1779-3 ISBN 978-981-92-1780-9 (eBook)
https://doi.org/10.1007/978-981-92-1780-9

This Springer imprint is published by the registered company Springer Nature Singapore Pte Ltd.
The registered company address is: 152 Beach Road, #21-01/04 Gateway East, Singapore 189721, Singapore

Dedicated to the researchers, educators, and innovators shaping a sustainable and human-centered technological future.

Foreword

This book brings together a collection of scholarly perspectives that illustrate how the transition from Industry 4.0 to Industry 5.0 represents not only a technological leap but a philosophical one placing human well-being, sustainability, and inclusivity at the heart of innovation. The editors present this volume as a reference for academicians, researchers, and practitioners seeking pathways toward responsible and human-centric industrial transformation.

Dr. Sanjeev Prashar
Professor
Indian Institute of Management
Raipur, India

Preface

The Fourth Industrial Revolution marked a transformative period of automation, algorithmic intelligence, and hyper-connectivity that reshaped industries, economies, and human interactions. Smart systems began to decide, predict, and optimize with a level of precision unimaginable only a few decades ago. Yet, amidst this remarkable acceleration, an important realization surfaced: innovation that sidelines human values risks creating efficiency without empathy and progress without purpose.

Industry 5.0 emerges as a philosophical evolution of Industry 4.0—restoring human creativity, compassion, and consciousness to the center of technological design. It does not reject automation or data; rather, it reframes them as instruments to elevate human potential and societal well-being. This edited volume, **Integrating Industry 5.0 and 4.0 for Technology and Humanity**, responds to this shift by exploring how technology, sustainability, and human-centricity intersect across diverse domains such as consumption, finance, education, workforce development, and social inclusion.

The **11 chapters** of this book are organized into **four thematic parts**, collectively narrating a journey from automation to awareness, and from intelligence to intuition.

Part I—Human-Technology Integration

This part examines how human agency, meaning-making, and improvisation remain essential in an increasingly institutionalized technological environment.

Dr. Guninder Pal Singh explores the interplay between institutionalized consumption and user improvisation, demonstrating how individuals negotiate, resist, and co-create digital routines. His work highlights that authentic innovation thrives when human agency—not technological determinism—guides system evolution.

Part II—Digital Finance, AI, and Intelligent Innovation

The part section brings together contemporary perspectives on how Artificial Intelligence and emerging digital infrastructures are reshaping financial ecosystems, advisory services, and economic behavior.

Anubhav Roy and Dr. Priyanka Singh foreground the role of human-centric AI in improving financial decision-making and enhancing the accessibility of advisory services.

Dr. Vaibhav Aggarwal, Dr. Mahender Yadav, Mr. Shashank Sharma, and Dr. Reepu articulate how algorithms are evolving into intelligent financial advisors, offering new possibilities for risk assessment, portfolio management, and investor empowerment.

Dr. Indira Bharadwaj and Dr. Reepu provide a comparative assessment of volatility dynamics in green and conventional cryptocurrencies, illustrating the tension between sustainability narratives and speculative forces.

Dr. Abha Gupta, Dr. Kamini Rai, and Mr. Manoj K. Mishra present a systematic mapping and bibliometric analysis of Non-Fungible Token (NFT) research, tracing its development and linking digital assets with principles of circular economy and sustainable digital innovation.

Together, these chapters illuminate how finance and technology converge to create intelligent, responsible, and resilient economic systems.

Part III—Digital Markets, Learning, and Workforce Behavior

This part shifts the focus to the behavioral, educational, and experiential dimensions of the digital era.

Sorabh Bajaj and **Dr. Mamta Sharma** develop a theoretical framework for web-based instruction, emphasizing the role of cognitive design, learner motivation, and human–machine interaction in effective digital pedagogy.

Choo Siao Ying, Dr. Low Mei Peng, and Dr. Har Wai Mun conceptualize well-being among gig workers through the Job Demands–Resources model, highlighting how flexibility and uncertainty shape the lived realities of digital labor markets.

These chapters remind us that digital markets are not merely economic exchanges—they are ecosystems where humans learn, aspire, adapt, and seek dignity.

Part IV—Sustainability, Inclusion, and Empathy-Driven Organizations

The final part connects technological evolution with ecological responsibility, social justice, and human well-being—core pillars of Industry 5.0.

Dr. Manidipa Das Gupta and Dr. Sumanta Majumder investigate how decent work conditions shape mental-health well-being and productivity in India's brick-field sector, underscoring the inseparable link between humane labor practices and organizational outcomes.

Vibhor Kataria and Dr. Tavleen Kaur examine transgender inclusion in white-collar employment, offering policy insights aligned with the United Nations Sustainable Development Goals and advocating for workplaces grounded in dignity and equality.

Ms. Shikha Shokeen and Ms. Alisha Khattar propose strategic pathways for achieving Net-Zero goals by integrating environmental stewardship with business competitiveness, cautioning against superficial greenwashing and advocating transformative sustainability.

Ms. Ankita Bansal and Dr. Anjali Vyas demonstrate how green, socially responsible, and sustainable HRM practices can catalyze organizational transformation, aligning people practices with ecological and social objectives to create holistic, innovation-driven HRM systems.

Collectively, these contributions illustrate that sustainable and inclusive practices are not peripheral obligations—they are engines of long-term innovation, resilience, and organizational authenticity.

Closing Reflections

Beyond documenting technological progress, this volume captures a deeper philosophical transition—from industries that compete for efficiency to ecosystems that collaborate for shared prosperity. It invites scholars, practitioners, and policymakers to reconsider what constitutes "success" in an age of smart machines: not domination over technology but partnership with it.

The editors hope that the reflections and frameworks presented herein will inspire future research and practice that value **empathy as much as efficiency**, **sustainability as much as scalability**, and **inclusion as much as intelligence**.

The spirit of Industry 5.0 is not about replacing humans with machines; it is about empowering humans *through* machines to build a future that is intelligent, inclusive, sustainable, and profoundly humane.

New Delhi, India — Dr. Bhajneet Kaur
Abu Dhabi, United Arab Emirates — Dr. Sudhir Rana
Noida, India — Dr. Sudhi Sharma

Acknowledgments

The editors express sincere gratitude to all contributing authors, reviewers, and institutional partners for their commitment and intellectual engagement. Appreciation is also extended to the *College of Business, Liwa University, Abu Dhabi*, UAE, *Rukmini Devi Institute of Advanced Studies, New Delhi, India,* and *Amity University, Noida, India* for supporting interdisciplinary collaboration.

Finally, the editors thank Springer Nature's editorial and production teams for their guidance and professionalism throughout the preparation of this volume.

Contents

About the Editors

Dr. Bhajneet Kaur is Associate Professor at the Rukmini Devi Institute of Advanced Studies (RDIAS), India, specializing in Business Analytics. Currently she is working as Area Chair (IT and Business Analytics). With over 12 years of experience in teaching and research, Dr. Kaur also holds a "Global Certificate in Data Science" from the International School of AI and Data Science.

She is Sought-After Resource Person and Industry Trainer, having been invited to conduct sessions in Faculty Development Programs (FDPs), Management Development Programs (MDPs), and workshops covering topics such as data analytics, research methodology, machine learning, AI, and related tools. Dr. Kaur's research has been recognized with accolades, including the Best Paper Award at the Information Security Risks 2018 International Conference. She has presented her work at numerous prestigious conferences, including AICAI-2019 (Dubai), IIM, AOM-2023 (Boston, USA), IEEE, and Springer conferences.

Her research is published in various international journals including Scopus, SCI, WoS, and ABDC. Dr. Kaur's expertise lies in data and business analytics, descriptive and predictive modeling, machine learning, business problem-solving, and research methodology. She is proficient in using data analysis tools such as R, Python, SPSS, Power BI, NVivo, and Smart PLS. Additionally, she has served as Guest Editor for the International Journal of Public Sector Performance Management (IJPSPM) published by Inderscience Publishers. e-mail: bhajneetahuja@gmail.com

Dr. Sudhir Rana believes in driving and motivating academics and research in such a way that it can be best utilized in an enthusiastic and dynamic environment to foster versatile personalities. Dr. Sudhir Rana is Faculty of marketing and strategy at College of Business, Liwa University, Abu Dhabi, UAE. He holds Ph.D. degree with the Ministry of Human Resource Development Scholarship from the Government of India and Post-Doctorate from Putra Business School (AACSB Accredited), Universiti of Putra, Malaysia. He has set high standards in academics and research. Dr. Rana has been associated with Fortune Institute of International Business and the Indian Institute of Management, Nagpur, India. He has delivered 350+ workshop sessions, keynote speeches, and webinars. His research area is on digital technologies, consumer behavior, and emerging markets. His research is published in the Journal of Knowledge Management, Journal of Business Research, Journal of Consumer Behavior, Journal of Strategic Marketing, Journal of Marketing Theory and Practice, TQM Journal, Journal of Global Marketing, International Journal of Emerging Markets, and Journal of Promotion Management among others. A few stories that he turned around include the progression of the journal FIIB Business Review as Editor-in-Chief in making the journal among most-preferred journal from Sage Publishing (achieved indexing and ranking in the Scopus, Web of Science, ABDC, and ABS). He launched as well as chaired the Doctorate Program (Executive/Fellow Program in Management—equivalent to Ph.D. approved by AICTE, an apex body Government of India), at Fortune Institute of International Business, India, as Founding Chairperson. In January 2021 he joined the College of Healthcare Management and Economics (CoHME), Gulf Medical University, UAE, and contributed to establishing CoHME as a stable constituent of the university in a turbulent time. He is experienced in various positions like Department Chair, Coordinator Corporate Internship Program, etc. He has been Speaker at various forums such as Academy of International Business (AIB) and Academy of International Business and Professional Management (AIBPM). He did many special issues and is presently working as EIC of FIIB Business Review (Sage), Co-EIC of

South Asian Journal of Marketing (Emerald), Series Editor of Review of Management Literature (Emerald), and Advances in Emerging Markets and Business Operations (Taylor and Francis). https://www.emerald.com/insight/publication/issn/2754-5865 e-mail: rana.sudheer21@gmail.com

Dr. Sudhi Sharma is presently working as a professor at Amity University, Noida, India. Prior to this she has worked as Program Chair and a core finance faculty at FIIB, New Delhi. Prior to joining FIIB, she has taught at various reputed B-Schools like BIMTECH, Apeejay, JK Group to name a few. She has an experience of more than one and half decade in research, industry training and academics. Her recent MDP trainings were focused on Financial Modelling, ESG Reporting, Financial Econometrics, Data Analytics in R, Data Visualisation on PowerBI and Advance MS-Excel and strategic planning for various companies from both public and private sector. She has conducted training programs for companies and ministry officials like Ministry of Tanzania, Walter Bushnell, Ministry of Textile, India. She has conducted EDPs and workshops for the delegates of RBS, Genpact, CII, and Ummeed Housing Finance. She has also handled training programs for Red Cross employees from Afghanistan. Her core interest lies in Financial analytics, Financial Modelling, Data Analytics, Predictive Modelling, etc. She has also published research papers in ABDC indexed journals. You can reach at some of her work with the link below:

Link: https://www.scopus.com/authid/detail.uri?authorId=57220130204

Apart from training industry professionals she has also trained faculty members from technology background on Data Visualisations and Predictive Models and has taken several Faculty Development Programs in house and other institutions. She always cherishes taking up newer topics in the area of Data Analytics and believes that innovative exposure can keep professionals on track with challenges in the corporate world. e-mail: sudhisharma1983@gmail.com

Abbreviations

ADF	Augmented Dickey Fuller Test
AI	Artificial Intelligence
AMOS	Analysis of Moment Structures
CSR	Corporate Social Responsibility
DEI	Diversity, Equity, and Inclusion
DeFi	Decentralized Finance
DPoS	Delegated Proof-of-Stake
EMH	Efficient Market Hypothesis
ESG	Environmental, Social, and Governance
ETF	Exchange-Traded Fund
GARCH	Generalized Autoregressive Conditional Heteroskedasticity
GHRM	Green Human Resource Management
HRM	Human Resource Management
I4.0	Industry 4.0
I5.0	Industry 5.0
ILO	International Labour Organization
IoT	Internet of Things
ISM	Interpretive Structural Modeling
JD–R	Job Demands–Resources Model
JS	Job Satisfaction
LGBTQIA+	Lesbian, Gay, Bisexual, Transgender, Queer/Questioning, Intersex, Asexual (+)
LQ	Life Quality
MICMAC	Cross-Impact Matrix Multiplication Applied to Classification
NFT	Non-Fungible Token
NIOSH	National Institute for Occupational Safety and Health
OSH	Occupational Safety and Health
PoS	Proof-of-Stake
PoW	Proof-of-Work
PRISMA	Preferred Reporting Items for Systematic Reviews and Meta-Analyses
RPA	Robotic Process Automation

SDG	Sustainable Development Goals
SEM	Structural Equation Modeling
SHRM	Sustainable Human Resource Management
SLR	Systematic Literature Review
SPSS	Statistical Package for the Social Sciences
SRHRM	Socially Responsible Human Resource Management
UN	United Nations
WASH	Water, Sanitation, and Hygiene
WBI	Web-Based Instruction
WEF	World Economic Forum
WHO	World Health Organization
WLQ	Work–Life Quality

Part I

Human–Technology Integration

Institutionalization in Consumption Meets Improvisation: A Case Exemplar Approach

1

Guninder Pal Singh

Abstract

Consumption does not arise from a single cause (e.g., a campaign), but rather from an interplay of multiple internal and external drivers, including psychological, social, cultural, affective, and cognitive factors. These factors are embedded in consumers' personalities, immediate surroundings, and situational contexts. Their relationship makes consumers' lives and decision-making a complex matter that even the most comprehensive marketing guidance can only partially anticipate. Although firms provide product guidance and brand narratives to help consumers navigate this complexity, it is during the usage episode that consumers improvise their practices. While improvisation is typical in consumption episodes, it is usually inadequately acknowledged. Using an interpretive paradigm, the paper employs case studies to elucidate how consumption practices lead to improvisations. We examine how consumer-led improvisation thrives outside the institutionalized consumption norms. We note that consumers oscillate between following and bending institutional guidance, and the previously untapped "value pockets" become available. A conceptual framework is proposed to illustrate the interconnectedness of institutionalized and improvised consumption. Thus, the framework sheds light on their mechanisms and outcomes.

Keywords

Consumer improvisation • Institutionalized consumption practices • Value creation • Technology platforms • Case study

G. P. Singh (✉)
Jindal Global Business School, O.P. Jindal Global University, Sonipat, Haryana, India
e-mail: Guninderp.singh@jgu.edu.in

B. Kaur et al. (eds.), *Integrating Industry 5.0 and 4.0 for Technology and Humanity*, Contributions to Environmental Sciences & Innovative Business Technology,
https://doi.org/10.1007/978-981-92-1780-9_1

1.1 Introduction

Consumption is a complex phenomenon that cannot be reduced to a linear managerial stimulus, such as advertising or an influencer campaign. How consumers respond to these stimuli depends on several factors like motives, values, reference groups, emotions, beliefs, evaluations, among many others. These factors cumulatively describe a person and their context, i.e., who they are, where they are, and what is happening. Overall, the joint influence of a person's psychological, social, cultural, affective, and cognitive states impacts their behavior and response (Arnould and Thompson 2005). Consumers do not always behave as the firm intends them to, despite its persuasion efforts and guidance. It is common to note consumers use a product in unintended ways, combining offerings across categories, creating new content formats, or repurposing a brand for a different symbolic project than originally envisioned. These deviations might seem anomalies, but should be seen through the lens of consumer improvisation. It serves as an explanatory phenomenon that clarifies why and how people behave outside of the patterns firms might anticipate. It is a set of spontaneous and creative behaviors. While these behaviors might use market cues, their nature and execution deviate from expected market/firm patterns (Coker et al. 2024). Thus, improvisation becomes a meaning-making and transformative process, not merely a behavioral deviation, wherein consumers' efforts instill meaning in consumption episodes that exceed routine outcomes, such as satisfaction, and also often generate higher-order consumer states, including empowerment and even transcendence (Fortezza et al. 2022). The entire sequence of improvisation is characterized by uncertainty in the situation.

It may not be possible for the marketing teams to factor in the uncertainty in their marketing efforts wherein they design strategies to deliver results (e.g., sales targets, campaign performance, ROI). To do so, they make an assumption that consumers' behavior can be predicted. This raises a paradox: market success hinges on accurately predicting consumer behavior, yet there is inherent uncertainty in consumption episodes. In this predictive paradigm, firms mobilize an array of marketing tools and frameworks, such as data analytics, consumer profiling systems, and multi-method market research programs, designed to infer, model, and anticipate consumer choices (Homburg et al. 2015). Repeated use of such analysis triggers a form of institutionalization wherein a common and generalizable understanding of the diverse human, sociocultural, and environmental elements gets created. However, the predictive techniques convert the heterogeneous consumer situations into homogenous variables and segments, resulting in a rather oversimplified view of the consumer. Using this consumer view, firms then shape several activities like product design, marketing strategies, and branding. The goal is to better align value propositions with the newly identified homogeneous consumers' needs, personalities, situations, and expectations. While this structure provides market efficiency, it constrains spontaneity in consumer behavior, by narrowing the scope for individual creativity and improvisation during consumption episodes (Rifkin et al. 2023).

Notably, consumers frequently find themselves navigating, often within the same episode of use, both firm-controlled institutionalized practices and naturally occurring improvisation in their consumption experiences. While institutionalized practices prescribe how products should be approached, evaluated, and used, the improvised practices are based on spontaneity, creativity, and experimentation. In that sense, consumer improvisation can be seen as an alternate narrative to firm-controlled institutionalized norms. Although both mechanisms coexist, their mutual interpenetration and the forms of tension and alignment that arise between them remain, as we argue here and show in the subsequent sections, underexplored in scholarly inquiries. Marketers aim to generalize and predict behavior, while consumers' delight might stem from a place of improvisation and unpredictability (Arnould and Thompson 2005). A fledgling strand of literature explores consumer improvisation (Coker et al. 2024; Demetry 2024), but such works largely tackle its definitional and scope-related discourse. Its interplay with institutionalized practices is lacking and remains imperative because users are clearly deploying them and its under-exploration keeps the improvisational practices in a black box. Thus, as consumers navigate these mechanisms simultaneously, looking at the interplay promises relevant insights. This paper examines how consumer improvisation interacts with institutionalized marketing practices (Lawrence and Suddaby 2006). Institutional theory seeks to recognize and understand how external social structures and institutional arrangements shape and constrain human action (DiMaggio and Powell 1983; Slimane et al. 2019). Institutional theory in consumption contexts posits that what individuals perceive as personal choices is inherently rooted in the norms and values institutionalized by firms.

Given this context, this study explores two questions: (1) What mechanisms enable consumers to improvise during consumption? (2) How does the interplay between institutionalized and improvised consumption lead to new forms of value? With the help of case exemplars, we demonstrate that innovation and new forms of value creation are possible outside of the institutionalized understanding of consumption through improvisation mechanisms (Scaraboto and Figueiredo 2022; Vargo et al. 2023). An inductive methodology is adopted as the intent is not to test the a priori relationships between constructs. Instead, it examines the extant and emerging relationships from a novel perspective and looks for underlying mechanisms that enable improvised consumption (Eisenhardt and Graebner 2007). We find mechanisms enabling improvisation: shared interests and intrinsic motives, peer-to-peer networks, creation of digital platform collectives, uncertainty and risks in the consumption episode, the ability of users to switch the capacity of interactions, and their ability to communicate and collaborate across geographies. In doing so, the paper provides substantial theoretical insights regarding consumption processes. It illuminates an interplay between institutionalized and improvised consumption, recognizes novel manifestations of consumer empowerment emanating through improvisation, demonstrates the emergence of brand and cultural meanings, and highlights the importance of technology platforms in enabling non-traditional behaviors.

We begin this paper by examining the literature on (1) improvisation, (2) consumer participation, and (3) the application of institutional theory in the context of consumption. Following an explanation of our inductive approach, data selection, and analysis procedures, we analyze case exemplars that illustrate our conceptual framework of new value creation through the interplay of improvised and institutionalized consumption. Finally, we discuss the theoretical and managerial implications of our findings.

1.2 Literature Review

1.2.1 Consumers and Improvisation

Improvisation is a common aspect of lived experiences, but it is not often discussed. Individuals often discuss applying several related concepts, such as creativity, spontaneity, serendipity, or unplanned action. Improvisation is a higher-order construct, cumulative of these related ones. In other words, it can be described as a process in which planning and execution converge in real time, allowing actors to respond effectively to unforeseen circumstances rather than executing pre-given instructions (Coker et al. 2024). While these related concepts have an interplay and can operate in a nested and meshed manner, spontaneity is considered an instrumental factor. It works as an interplay of cognition, behavior, and context, enabling individuals to produce context-specific solutions that materialize as improvisational episodes (Rifkin et al. 2023; Weick 1998). In a way, from a processual standpoint, actors' responses exhibit an odd juxtaposition of surprise, creativity, and efficacy in the face of unforeseen circumstances, allowing for the spontaneous enactment of creative actions without prior planning or anticipation. Improvisation emerges as an outcome of this process, in which individuals enact their past experiences, know-how, and situational factors into actions (Mannucci et al. 2021). Improvisation is a necessary attribute in the service industry. For example, front-line staff at airports, restaurants, banks, and events are faced with diverse, often odd, and complex situations while interacting with customers. These unpredictable complexities emerge from customers and/or situations of the context. Sometimes, improvising is required for better services, and not problem-solving. For example, doctors and educators observe and think on the fly while handling queries and problems. In all these performative domains, professionals move between prepared material and in-the-moment adaptation (Weick 1998).

From a consumer's standpoint, improvisation typically revolves around the product as the focal point. All creativity and adaptations utilize the product as the vehicle, wherein consumers assume the driver's seat and chart out their pathway based on their needs and in situ assessment of the situation, ultimately aiming to maximize their utility and problem-solving (Coker et al. 2024). This behavior is prevalent as consumers seek personalized and fulfilling consumption experiences (Vargo et al. 2023). Consumers are driven to improvise when they encounter gaps between what is available and what they require, prompting them to modify standardized offerings to suit their context (Scaraboto and Figueiredo 2022). Rather than merely accepting, these behaviors reflect consumers' active role in. Here, the motives of improvisation include unmet needs from typical marketplace guidance, customization to suit a context, or spontaneous problem-solving using the product, wherein consumers go beyond the standardized instructions and offerings and shape their product experiences using socio-affective-cognitive ingenuity and resourcefulness (Demetry 2024). At the same time, it would be reductive to view improvisation solely through an instrumental, need-fulfillment lens. Furthermore, engaging in improvisational activities gives consumers a sense of empowerment and control over their consumption experiences, as they move from passive uptake to active re-making. By doing so, consumers experience themselves as agents who can intervene in, rather than simply submit to, the marketplace's structures. Ultimately, consumers can express their individuality and achieve a sense of uniqueness by modifying standardized marketplace offerings to create personalized experiences that more closely reflect their identities and preferences. In the process, consumers interact with other users in their personal circles or other individuals on social media, gaining new product usage insights. Social media and personalization engines play a crucial role in connecting interests with similar interests, enabling interactions and exchange between them. Thus, at the core, consumer improvisation is not only an individualized phenomenon, but it also factors in the role of social and relational exchange as well.

1.2.2 Consumer Participation

Consumers are the source of inspiration for designing marketplace activities. According to historical views, consumers had the sole function of consumption. With time, consumers also engage in marketplace activities beyond consumption. They now partake in marketing, suggesting new product designs, promotional campaigns, price determination, among other marketplace activities. Consumers interact with firms, regulators, and other fellow consumers. These interactions may be direct or mediated, one-to-one, or one-to-many, transactional or relational. All such interactions require involvement and participation from consumers. Consumer-to-consumer interactions are a substantial force of momentum affecting brand perceptions and consumer sentiments. These interactions aren't typically part of the roles attributed to consumers as per the consumer-firm dichotomy. Gradually, due to the emergence of social media and digital infrastructure, consumers engage in user-generated content that affects all marketing and strategic decisions for a firm (Muñiz and O'Guinn 2001; Bleier et al. 2024). The power of virality is not limited to consumers, implying that the impact of non-users' perception on image and perception is just as powerful as that of users. Clearly, consumers' participation has witnessed exponential growth. Thus, consumers (whether users or non-users) are not merely to be seen as buyers but as integral participants in the value-creation process. Consumers engage with brands in all stages of their purchase journey—before, during and after the purchase decision. These interactions happen with different market and non-market actors, depending on the channel of communication and the stage of the consumer journey. For example, potential users engage with informational content on social media in pre-purchase stages, as the information helps them navigate their questions. However, engagement mode shifts to entertainment and emotional appeals in the post-purchase phase, wherein the interactional value is driven by experiences and assessments instead of exploration and examination. In these settings, we see value emerges as a function of interactions. In a physical setting, the role of front-line employees is more; hence, it is important for them to use affective states like empathy and humor even in the pre-purchase state. In all these cases, interactions generate different forms of value and use different mechanisms, with the product remaining constant. These observations suggest that the value is not solely inherent in the product, but rather it emerges from a network of interactions at various stages of the journey and varies across different channels.

Consumer participation manifests through various mechanisms. Consumer psychology plays a crucial role in encouraging people to engage with the brand beyond consumption goals. Essentially, what consumers feel and think about themselves in the process guides their motivations to participate. In settings where consumers co-produce value on top of the raw material provided, they attribute the process to themselves. When the outcome quality is good, consumers form a positive association with the brand and product, thereby creating a loop of future brand engagement and participation. In a sense, consumer preferences and expectations are nested in the process of value creation (Cova and Cova 2002; da Fonseca et al. 2023). Consumers also experience empowerment and control in the consumption process (Fortezza et al. 2022; Füller et al. 2009). From the firms' perspective, consumer insights and ideas fuel product development, communication, and brand building through UGC; all these perspectives might otherwise remain outside the organizational horizon (Füller et al. 2009). Consumer participation can also impact the product domain. For example, many firms have begun relying on consumers to seek product suggestions and select the best alternatives (Moorman and Miner 1998). Similarly, new business models built on the sharing economy demonstrate that the impact of resource sharing can create a multiplier effect of growth, simultaneously disrupting existing markets and generating new opportunities. Such models benefit from active consumer participation in distributed consumption networks (Botsman and Rogers 2010; Bleier et al. 2024).

As we can see, the benefits of consumer participation are immense and multidimensional. The participatory mechanisms are often institutionalized, governed, and direct consumers' work (Beverland et al. 2024) as they engage in the marketplace. This institutionalization of consumer work does not fully account for or accommodate several unconventional, spontaneous, and unanticipated behaviors, thereby curtailing the impact of consumers' participation in the process. Such unplanned behaviors often result in new sociocultural and market-centric outcomes.

1.2.3 Institutionalization of Consumption Practices

According to institutional theory, markets are institutions comprising social structures, norms, rules, and understandings (Lawrence and Suddaby 2006). Institutional theory contends that firms actively seek market legitimacy by adhering to societal norms, values, and expectations (Slimane et al. 2019). On the other hand, firms devise commerce, social, and community logics to trigger consumer engagement with their value propositions (Scaraboto and Fischer 2013; da Fonseca et al. 2023). Institutionalizing consumers' consumption habits corresponding to a firm's desired outcome creates unique brand positioning. Overall, firms mobilize human desires and needs in a manner that consumer action becomes a legitimization of the firm's products and positioning. In the process, to expedite the legitimization process, conformity in bigger consumer segments becomes efficient and rewarding decisions for the firm (Zanette and Scaraboto 2019). Understanding consumer preferences and placing the products closer to them increases the chances of choice. In a way, institutionalizing aims to code users' preferences, so the choice becomes an inevitability. Thus, firm positions are often institutionalized goals framed as consumer preferences. It becomes apparent that while such institutional arrangements and firm-authored templates bring a degree of consistency to consumers' actions, they simultaneously operate as devices that can evoke, and at times intensify, conformity to predefined behavioral and symbolic norms. Consumers respond to and navigate this delicate balance between conforming and asserting their individuality through market-mediated identity work in consumption (Dolbec and Fischer 2015). Their response may be planned and orchestrated or spontaneous and unanticipated (Scarboto and Figueiredo 2022). In addition, consumers sometimes respond to institutional pressures not by subtle reconfiguration but by adopting explicitly resistant or counter-hegemonic positions, mobilizing sanctioned consumption practices such as minimal consumption, anti-consumption, or deliberate refusal as ways of distancing themselves from dominant market logics (Wilson and Bellezza 2022). In the case of marginalized and excluded individuals, entrepreneurship becomes an act of curating legitimacy in the mainstream market when existing consumer segmentation does not recognize them (Scaraboto and Fischer 2013). In cases where consumers' specific functional needs are not met, they may exhibit ingenuity and creativity in altering their normal consumption patterns by modifying standard products, which can even result in the creation of alternative marketplaces (Martin and Schouten 2014; Sacroboto and Fischer 2013). The above responses often involve consumer work (Beverland et al. 2024) to actively counter and resist existing choices, requiring a thought-out and measured approach to creating new marketplace alternatives. On the other hand, some consumer responses may be unanticipated, novel, and creative, without intending to express any resistance. In these cases, improvisation may occur due to an inherent, intrinsic motivation to experiment with the product without counter motives. Inherent passion and attachment toward the product might trigger consumer experimentation. Such creative improvisations often give rise to informal and small-scale subcultures and social collectives among individuals who share similar motives and attachment objects.

In that sense, improvisation includes neutrally alternative and counter-narratives to firm-created consumption practices and norms. They often occur in environments outside the purview of firm, institutionally established practices.

Clearly, both institutionalized and improvised consumption offer benefits to the market and consumers. Creating practices and ecosystems encouraging complementarity between the two stands to enhance overall market outcomes for diverse stakeholders.

1.3 Research Methodology

1.3.1 Research Approach

This inquiry adopts a constructivist–interpretivist epistemology in which social reality is treated as emergent and negotiated. Consumer improvisation is therefore examined as a situated process rather than a set of stable variables. We employ case study research for interpretive sense-making—the fourth of Welch et al.'s (2020) theorizing logics—because we aim to develop new conceptual insight, not to test deterministic hypotheses. Our methodology is designed to explore the mechanisms through which consumers improvise and how these behaviors interact with institutionalized consumption to generate new forms of value. By considering several situational factors in different contexts, case exemplars effectively conceptually interpret a phenomenon emanating from practice (Kolyperas et al. 2019; Welch et al. 2020). Following past works (Kolyperas et al. 2019) using a similar approach, we base our analysis on Welch et al.'s (2020) guidance on using cases as interpretive tools. Broadly, the four ways in which case studies can be used to theorize: (1) as inductive tools to test hypotheses; (2) as natural experiments that explain relationships in business settings; (3) as elements of a broader contextual epistemology providing a culturally grounded understanding of causal correlations in business relations; and (4) as tools for interpretive sense-making. Given that the goal of this study is to interpret and reflect, we evoke only the fourth perspective as it aligns with our aim.

1.3.2 Case Context and Analysis

We selected three focal cases: GoPro, MyFootballClub, and the Fan Controlled Football League (FCFL), because they demonstrated substantial marketing outcomes from consumer-induced improvisation practices (Kolyperas et al. 2019). Each case exemplar denotes elements of improvisation and participation, thus becoming relevant sites for this examination. We deliberately juxtapose a user-generated content platform (GoPro) with two digitally mediated fan-governance initiatives operating in sport (MyFootballClub and FCFL), to configure what Stake (2005) describes as a maximum-variation case set. Our prolonged (10+ years) engagement with these domains enhances our sensitivity to context-specific nuances and allows us to recognize patterns that might be less visible to more distanced analysts.

Our analysis of the three exemplars draws on a corpus of secondary and archival data comprising company filings, academic articles, management interviews, and media reports. To retain the meaning and richness of each exemplar, we first produced "thick descriptions" (Stake 2005) for each case separately and sequentially. We engaged in a continuous back-and-forth between emergent theoretical insights, case-specific evidence, and the broader institutional and market contexts. Following Eisenhardt and Graebner (2007), we seek to enhance the internal validity and theoretical robustness of our claims. To further strengthen the credibility of our interpretations, we engaged in peer debriefing with two senior faculty members specializing in marketing. The interpretations were systematically examined collectively, with special emphasis given to blind spots and alternative explanations.

1.4 Analysis and Discussion

In this section, we demonstrate how certain firms foster and sustain innovation and loyalty by engaging consumers in a non-institutionalized capacity. In the three examples below (Table 1.1), consumers engage with the core offering in ways not often predicted or anticipated by the respective firms. In that sense, consumers deploy innovation and improvisation while using the product/service.

1.4.1 Integrated Framework

We theorize a framework from the case exemplar observations, showing that market exchange is enhanced when institutionalized and improvised forms of consumption interact and coexist. Institutional consumption practices are anchored in firm-controlled systems, and improvised consumption emerges when consumers depart from and improvise the firm's instructions. The transition between institutionalized and improvised behaviors enables consumers to generate value independently of firm control, and unlocks untapped value pockets—unconventional behaviors, user-based ownership, micro-entrepreneurship, and the emergence of new cultural meanings. Institutionalized consumption, standardized and routinized consumer behavior guided by the firm, is contrasted with improvised consumption, which occurs when consumers engage in extemporaneous behaviors that were not anticipated. Let us examine the specific mechanisms at play:

Improvisation Mechanisms

These mechanisms demonstrate how consumers navigate and sometimes transcend institutionalized practices, depicting the improvisational behaviors.

Across all three exemplars, *actor role-switching* is a salient pattern consumers depict. This takes after the foundations of the service-dominant logic, the paradigm

Table 1.1 Case exemplars and analysis

Firm/initiative and description	Consumer improvisation elements
Firm/initiative: GoPro Title: Novel fan experience fueling further innovation GoPro, established in 2002, is an action camera manufacturing company. Adventure sports enthusiasts primarily use its cameras to record their own experiences. The device's uniqueness lies in enabling the fans to record from any angle without external help. These cameras enable fans to enhance their sporting experience, creatively build their narrative, showcase their skills to their peer groups, and engagingly share information with fellow fans The videos made by the fans and shared by them on social media have become the bedrock of success for the company. Videos recorded using GoPro cameras act as fan testimonials and guerrilla advertising for GoPro products. As of June 2020, GoPro had 10.8 million fans on Facebook, 9.17 million subscribers on YouTube, 17.2 million fans on Instagram, and 2.1 followers on Twitter Apart from the direct benefit to the firm, such videos increase the popularity and participation in such sports Sample Sources: (Jarzemsky 2014; Lapowsky 2014)	• Users improvise unconventional ways to use GoPro cameras, spontaneously capturing unique perspectives and risky activities hands-free. Users employ cameras in activities beyond extreme sports, such as cooking tutorials, travel vlogs, or pet adventures, demonstrating the product's versatility • Users collectively collaborate by improvising in editing and compiling footage to create engaging and viral narratives that reflect their personal experiences and identities • Users' videos often go viral, resulting in organic advertising that significantly enhances GoPro's brand visibility and achieves positive marketplace results

(continued)

Table 1.1 (continued)

Firm/initiative and description	Consumer improvisation elements
Firm/initiative: MyFootballClub Title: Commercial outcomes of inclusion of fans in governance The example of MyFootballClub demonstrates non-traditional behavior by fans due to their heightened emotional connection with the sport. Fans on MyFootballClub, an internet community of soccer fans, raised money (£35 per fan) to purchase an operational soccer club in the Conference League in the UK. The fans-cum-owners would have no rights to any economic return for their investment; however, they had a voice in decisions such as manager selection and player transfers. The MyFootballClub community collected over £500,000 from fans from over 120 countries within ten days of the start of the campaign. With over 95% of members voting a green signal, the Ebbsfleet United F.C. was purchased for a sum of £635,000. MyFootballClub became the first case of a crowdfunded community, becoming the owner of a fully functional competitive sports club. The club's ownership was gradually changed, and two-thirds of the stake was sold to the Fleet Trust Sample Sources: (Cole 2017; Tomczak and Brem 2013)	• Fan-created platform united supporters globally, enabling collaboration, communication, and ownership • Used forums, social media, and email to discuss strategies, share updates, and engage with other members and club officials • Created a unique collective identity by jointly experiencing the ups and downs of club ownership and management • Fans improvised within sport management structures, utilizing technology and collaborative strategies to unlock new value forms

(continued)

Table 1.1 (continued)

Firm/initiative and description	Consumer improvisation elements
Firm/activity: Fan controlled football League Title: Fans as decision-makers Fan Controlled Football League (FCFL), originally started as Project Fanchise in 2015, is an American Football league that involves fans in several strategic and tactical decisions, ranging from coach or player selection to team nickname selection to picking plays during the games. The league is yet to host its first entire season, which will have four teams, and the matches will be played in a technology-powered stadium in Las Vegas. Previously, the franchise project involved a single team (Salt Lake Screaming Eagles) participating in the 2017 season of the Indoor Football League in the US. Due to high management costs, all four owners (all with NFL and technology backgrounds) of FCFL decided to pull the team out of the tournament and launch an entire league instead In the league, each team will share coaching and support staff while maintaining a separate roster of players. Fans will be involved in all these related selections and other decisions via the league's mobile app. The fans get to watch the match on the twitch app, which has gamified interactions with the fans. The app seeks fans' input on play combinations transmitted to the coach, who finally relays the message to the on-field players The format is conducive to wooing digital fans, regardless of location. During the 2017 season, the Salt Lake Screaming Eagles team had more than a million fans for all games combined, with fans representing over 100 countries. Even before the COVID-19 pandemic affected the suspension of global sports, FCFL had contracted several famous personalities, such as boxing icon Mike Tyson and several former NFL players, as team owners. Due to restrictions on spectatorship in stadiums, a league like FCFL offers an alternate way to keep the fans engaged Sample Sources: Flynn (2016)	• Fans transitioned from consumers to decision-makers, enhancing involvement and accountability • Fans watch matches on twitch and participate in game decision with gamified interactions • FCFL's unconventional inclusion of fans in non-consumptive roles embraces spontaneity and risk, enhancing engagement through collective decision-making

Source Created by the author(s)

to popularize the notion of consumers being active stakeholders in the value-creation process. In each of the three cases, users oscillate between behaving as traditional consumers and as SDL-esque consumers. While donning the hat of a traditional consumer, users follow the directives given by the respective marketer. In the case of GoPro, users would see how-to videos, purchase the camera, and use it for filming their hobby activities. But consumers do not stop at that point; they enter the terrain of adventure and experimentation. Their experimentation does not end in their adventures, but it transcends and gets seen in how they share those videos with the world. That's the inflection point for the GoPro's consumers who become creators and marketing agents. Enabled by the firm, perhaps, not planned. Such oscillations are not rare; rather, they are now a feature of how users engage with GoPro, the product and the brand. In this trajectory, sharing with the outside world is a key activity. It emerges from a deep interest and passion for the activity, as well as a desire to communicate this to other fellow passionate individuals. This, the role transformations is fueled by a *shared passion* that keeps the improvisation engine revved up for more innovative and passionate behavioral patterns. In the case of GoPro, users' experiences are folded back into their identity-work, makes the camera a performative device, a portable identity-extension technology through which self-representation, adventure documentation, and brand storytelling are braided together. In the case of MyFootballClub and FCFL, as is often the case in sports, fans have a strong emotional connection with the sporting object (team, franchise, or sport). This shared passion and fanhood becomes results in them co-owning, managing, and governing key sporting decisions. In general, any act of proactive doing requires risk and outcome assessments before taking decisions. The nature of all three examples is such that users go through deep uncertainty in their relationship with sports (MyFootballClub), games (FCFL), and adventure sports (GoPro). The nature of lived uncertainty might be different in each of the three contexts; it nevertheless continues to be a common fabric. Sharing a passion is special; expressing the same can be liberating. The developments in digital technologies and the social media world have created an always-on, hyperconnected world, running deep on *peer-to-peer interactions.* The dimensions of these interactions are not just information, it also includes money, emotion, and identities, thus elevating the exchange as a higher-order construct, as can be observed in each of the three examples. Specifically, the MyFootballClub demonstrates power of monetary transactions, whereas GoPro and FCFL are unique cases for affective and identity exchanges. The key characteristic of social media is the creator engine it has created. Accordingly, the social media platforms have given rise to micro-influencers, who create DIY and how-to content, expanding the utility and possibilities of products (Kaplan and Haenlein 2011). The currency in the creator economy is *authenticity.* Specifically in the case of GoPro, several users have become influencers, and monetized their passion and love for their respective activity. Riding on authenticity, several subcultures get formed, becoming a place to share expertise and knowledge. The pattern is common in all three examples. Users today are fragmented across the globe, yet they remain connected, thanks to the always-on digital infrastructure of hardware and software. Despite geographical distance, *spatial connectedness* is salient among users in general. Otherwise low in number, sparsely spread consumers may not seem significant, but when connected and combined, their collective power outweighs the sum of their individual potential. In particular, MyFootballClub demonstrates the disproportionate impact of a globally dispersed fan base pooling nominal amounts to acquire an entire club. For FCFL, fans came together to govern and manage game decisions. Overall, both examples show a unique way of legitimizing a seemingly low-impact variable, in this case, spread-out users, when they are gathered together. The marketplace impact of this cohesion and coalescing of fans passions and commitments to their sporting objects gets a new meaning.

The Institutional and Improvised Consumption Interplay
Marketplace activity is not a unitary process but a composite arrangement in which firm-driven environments coexist with consumer-driven environments. Interestingly and paradoxically, consumers draw on institutional resources as they improvise and outgrow the institutionally guided behaviors. For example, GoPro consumers initially enter the system as relatively passive users, learning to use the camera and basic filming techniques by following guidelines provided by the brand itself. As users become proficient in the art of filming and posting their content on social media, they gather confidence and establish their digital identities and expertise. This creates a metaphoric loop wherein the master trains the disciple to outgrow him/her. For GoPro, user-generated campaigns are a key factor in its success. They demonstrate that the company created an appropriate balance between user instructions and improvisations. From a process perspective, GoPro's campaign architecture can be read as a deliberate attempt to organize and channel emergent creativity. Campaigns thus provide a structured opportunity for consumers to engage with the brand's marketing apparatus, comprising contest parameters and participation formats, creating an organized environment for content submission that is both structured and permissive. While users retain substantial creative freedom in capturing and editing their experiences, in choosing locations, narratives, risk levels, and aesthetic styles, GoPro supplies a distribution infrastructure, including official social media channels and the GoPro website, through which this content can be shared and ranked. It creates a consumer-space in which participation is guided but not fully determined. From a consumer's perspective, the 15 everyday users are turned into episodic co-marketers whose contributions are visible, comparable, and curatable.

Similarly, in the cases of MyFootballClub and FCFL, management provides fans with guidance, enabling them to become independent participants in the overall ecosystem. Specifically, fans use mobile phones and web portals, and they are given training, guidance, and tutorials on how to make game decisions and team selection calls before the league commences. The same firm provided training and digital infrastructure that allows users to rally around and form fan-momentum to get to the point of critical mass, wherein their voice holds value and substance.

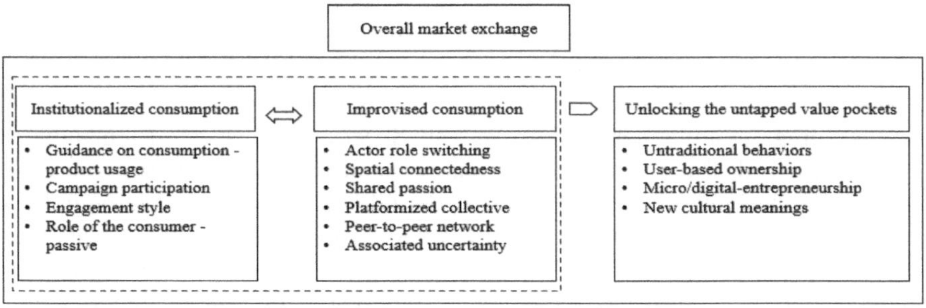

Fig. 1.1 Enhanced market exchange framework *Source* Created by the author(s)

Considered together, this dual impact of institutional and improvised consumption mechanisms opens up new value pockets. In traditional views, firms engineer value through their effort and assign a role to consumers for value capture. In the new consumer space envisioned through this new lens, users do not passively spectate as firms create value while waiting in the background, acting as legitimizing agents for the firms to capture value. In this new space, consumers become creators of value. Understandably, the new forms of value aren't within the existing dictionary and vocabulary structures of the firm. But these new behaviors indeed broaden the overall pie of total value available (Fig. 1.1).

1.5 Theoretical Implications

Firstly, and most notably, this paper has extended the understanding of consumption processes by providing an alternate and augmenting perspective to the conventional view, which presents consumption behaviors as adhering to an institutionalized template. It extends the extant understanding of consumption by demonstrating how improvised consumption allows consumers to create and capture value separate from, alongside, and sometimes at odds with, firm-level control.

Contrary to the conventional view of consumer empowerment that largely revolves around consumers' ability to make choices from alternatives given by firms (Martin and Schouten 2014; Scaraboto and Figueiredo 2022), this study introduces an extended and global understanding. Through these improvised acts, consumers create a set-up of empowerment that is at once social, cultural, and personal. We propose that any view of empowerment rooted in individualizing choice through firm-defined assortments must be marked as inadequate, and it is the platform-mediated collectives described in the above sections that enable a fundamentally different form of empowerment to come into focus. In the collective settings, consumers engage in dynamic bargaining over what roles they play and how much say they have or want in the market. Crucially, consumers improvise within, outside and occasionally outside the institutional complexes to bricolage their own experiences. In this regard, improvisational consumption practices are those that outgrow the value absorption perspective toward the one involving the creation of new forms of value.

Emerging from this context, consumer improvisation *tangibilizes* brand and cultural meanings in a uniquely distinctive way. Specifically, instead of remaining at the recipient-end of the value propositions, consumers integrate brands into their ongoing life narratives and stories, making brand meanings tangible through lived episodes and situated performances. Consumers materialize brand values and cultural meanings through tangible outputs such as user-generated content, modified products, derivative artifacts, or communal activities, with these outputs functioning as material instantiations of otherwise abstract concepts. These improvisational outputs, which circulate within and across consumer collectives, thus operate as vehicles for the ongoing re-enactment and re-articulation of brand and cultural meanings.

1.6 Managerial Implications

We observe two prominent managerial implications:

The paper's application of improvised consumer behaviors, which are typically treated as noise at the margins of planned marketing systems, offers a set of managerial implications that are less about imposing additional control and more about learning to work productively with contingency. Firms and marketers are advised to acknowledge the limitations of preconceived templates and fixed models of consumer behaviors—the ones that presuppose stable response patterns and linear funnels. Rather than giving into their pre held biases and notions, marketers ought to live the philosophy of consumer-centricity. One of the discerning advice that companies should take to heart is acceptance that you can't completely predict the pathways consumers tread. Rather than always needing to understand and control coming interactions, marketers can intentionally leave some pockets of indeterminacy while creating boundaries of observation and post hoc explanation. In these indeterminate zones of unpredictable behaviors, consumers are given freedom as a resource to explore, rather instructions to follow.

Through this work, we observe a shift in managerial focus, wherein consumer interactions are understood not not merely an opportunity for need recognition and fulfillment; but also as a window into consumer-led innovation and empowerment. As people spend more time on social media, it becomes a channel for communication, but it also has the potential to initiate numerous non-product-related conversations. That's where the seeds of user-innovation and improvisation are sown. Due to the rise in the content economy, individuals are becoming social media celebrities, having authentic and relatable fan followings, making it imperative for brands to consider this space to embed brand stories through UGC. This will create collaborative environments for forms to create value propositions laden with improvisation possibilities, suited to both physical and virtual worlds.

References

Arnould EJ, Thompson CJ (2005) Consumer culture theory (CCT): twenty years of research. J Consum Res 31(4):868–882

Beverland MB, Fernandez KV, Eckhardt GM (2024) Consumer work and Agency in the Analog Revival. J Consum Res:719

Bleier A, Fossen BL, Shapira M (2024) On the role of social media platforms in the creator economy. Int J Res Mark 41(3):411–426

Botsman R, Rogers R (2010). What's mine is yours. The rise of collaborative consumption, vol 1

Coker KK, Zeiss JG, Albinsson PA (2024) The consumer on stage: toward a typology of improvisation in consumption contexts. J Bus Res 170:114305

Cole S (2017, Oct 26) What happened to MyFootballClub—the club where fans decided everything? Retrieved-The Guardian. http://www.theguardian.com/football/2017/oct/26/what-happened-to-myfootballclub-ebbsfleet-united

Cova B, Cova V (2002) Tribal marketing: the tribalisation of society and its impact on the conduct of marketing. Eur J Mark 36(5/6):595–620

da Fonseca ALA, Chimenti P, Campos RD (2023) 'Take my advice': entrepreneurial consumers and the ecosystemic logics of digital platforms. Technol Forecast Soc Chang 193:122601

Demetry D (2024) Embrace the unexpected: how organizations Foster participatory improvisation with customers. Organ Sci. https://doi.org/10.1287/orsc.2023.17551

DiMaggio PJ, Powell WW (1983) The iron cage revisited: institutional isomorphism and collective rationality in organizational fields. Am Sociol Rev 48(2):147–160

Dolbec PY, Fischer E (2015) Refashioning a field? Connected consumers and institutional dynamics in markets. J Consum Res 41(6):1447–1468

Eisenhardt KM, Graebner ME (2007) Theory building from cases: opportunities and challenges. Acad Manag J 50(1):25–32

Fan Controlled Football League (n.d.) The League. Retrieved from https://www.fcfl.io/the-league

Flynn E (2016) Project FANchise: behind the fan-run football team. Sports Illustrated. Retrieved: https://www.si.com/ifl/2016/06/07/project-fanchise-indoor-football-league-fans

Fortezza F, Figueiredo B, Scaraboto D, Del Chiappa G (2022) Managing multiple logics to facilitate consumer transformation. J Bus Res 144:377–390

Füller J, Mühlbacher H, Matzler K, Jawecki G (2009) consumer empowerment through internet-based co-creation. J Manag Inf Syst 26(3):71–102

Homburg C, Schwemmle M, Kuehnl C (2015) New product design: concept, measurement, and consequences. J Mark 79(3):41–56

Jarzemsky M (2014, June 24) GoPro brings rare gadget deal to hot IPO market. Wall Street Journal. Retrieved: https://www.wsj.com/articles/gopro-brings-rare-gadget-deal-to-hot-ipo-market-1403636467

Kaplan AM, Haenlein M (2011) The early bird catches the news: nine things you should know about micro-blogging. Bus Horiz 54(2):105–113

Kolyperas D, Maglaras G, Sparks L (2019) Sport fans' roles in value co-creation. Eur Sport Manag Q 19(2):201–220

Lapowsky I (2014, June 26) Why GoPro's success isn't really about the cameras. Wired Retrieved: https://www.wired.com/2014/06/gopro/

Lawrence TB, Suddaby R (2006) Institutions and institutional work, vol 2. The sage handbook of organization studies, pp 215–254

Mannucci PV, Orazi DC, de Valck K (2021) Developing improvisation skills: the influence of individual orientations. Adm Sci Q 66(3):612–658

Martin DM, Schouten JW (2014) Consumption-driven market emergence. J Consum Res 40(5):855–870

Moorman C, Miner AS (1998) The convergence of planning and execution: improvisation in new product development. J Mark 62(3):1–20

Muniz AM, O'guinn TC (2001) Brand community. J Consum Res 27(4):412–432

MyFootballClub (n.d.) About. Retrieved from https://myfootballclub.co.uk/about/
Rifkin JR, Du KM, Cutright KM (2023) The preference for spontaneity in entertainment. J Consum Res 50(3):597–616
Scaraboto D, Figueiredo B (2022) How consumer orchestration work creates value in the sharing economy. J Mark 86(2):29–47
Scaraboto D, Fischer E (2013) Frustrated fatshionistas: an institutional theory perspective on consumer quests for greater choice in mainstream markets. J Consum Res 39(6):1234–1257
Slimane KB, Chaney D, Humphreys A, Leca B (2019) Bringing institutional theory to marketing: taking stock and future research directions. J Bus Res 105:389–394
Stake RE (2005) Qualitative case studies. In: Handbook of qualitative research, 3rd edn. Sage publications
Tomczak A, Brem A (2013) A conceptualized investment model of crowdfunding. Ventur Cap 15(4):335–359
Vargo SL, Peters L, Kjellberg H, Koskela-Huotari K, Nenonen S, Polese F, Sarno D, Vaughan C (2023) Emergence in marketing: an institutional and ecosystem framework. J Acad Mark Sci 51(1):2–22
Weick KE (1998) Introductory essay—improvisation as a mindset for organizational analysis. Organ Sci 9(5):543–555
Welch C, Piekkari R, Plakoyiannaki E, Paavilainen-Mäntymäki E (2020) Theorising from case studies: towards a pluralist future for international business research. In: Research methods in international business, pp 171–220
Wilson AV, Bellezza S (2022) Consumer minimalism. J Consum Res 48(5):796–816
Zanette MC, Scaraboto D (2019) "To Spanx or not to Spanx": how objects that carry contradictory institutional logics trigger identity conflict for consumers. J Bus Res 105:443–453

Part II

Digital Finance, AI and Innovation

Human-Centric AI in Financial Management: Aligning Digital Transformation with Industry 5.0 Principles

2

Anubhav Roy and Priyanka Singh

Abstract

Artificial intelligence is changing the manner in which we handle finances. Besides automation with artificial intelligence and predictive analytics and real-time decision-making, most companies are also starting to transition to a sustainable model (Industry 5.0). This is why AI has to be introduced in the financial systems in an ethical manner with a systemic and strategic approach. This research paper will look at how artificial intelligence can enhance financial processing including evaluating the risk, identifying fraudulent behaviour and also strategic predictions. It will also assess the ethical consequences of integrating artificial intelligence into financial processing, both the organisational as well as the technical consequences. According to this research and the existing literature, the research will show the growing role that AI is playing in the contribution of employees, the capacity of financial institutions to perform resilience and the degree of interaction that customers are having with their financial services providers. The research will also touch upon the issues of providing the privacy of the data utilised by AI, the clarification of the decisions made by AI and training financial professionals in AI. Lastly, the research will give recommendations on how to build a framework of the adoption of AI in financial management using the integration of digital transformation practices and Industry 5.0 principles. These results will be incorporated into the more general scholarly debate concerning sustainable innovations that will facilitate

A. Roy (✉)
Department of Management Studies, Institute of Information Technology and Management (affiliated to GGSIPU), Delhi, India
e-mail: anubhav.roy@iitmipu.ac.in

P. Singh
Department of Commerce, Maharaja Surajmal Institute (affiliated to GGSIPU), Delhi, India
e-mail: priyanka@msijanakpuri.com

B. Kaur et al. (eds.), *Integrating Industry 5.0 and 4.0 for Technology and Humanity*, Contributions to Environmental Sciences & Innovative Business Technology,
https://doi.org/10.1007/978-981-92-1780-9_2

the close of the chasm between technological innovations and human values in the finance industry. In general, this paper will add to the existing literature by developing a highly considered conceptual framework that will integrate the concepts of Industry 5.0 in the context of AI-based financial management. Also, the research will provide new empirical data on a survey completed on Indian financial professionals on their experience in the adoption of AI in the finance industry; this is, they will talk about their adoption behaviour, training requirements, and their attitude towards the effectiveness of AI in the financial sector.

Keywords

Artificial intelligence • Financial management • Industry 5.0 • Digital transformation • Sustainable innovations

2.1 Introduction

Artificial intelligence is becoming a robust tool for redefining financial management, reforming the way businesses have operated to date and enabling ever-informed and data-driven decisions. Banks and financial institutions all over the world are embracing AI-based tools to make complex operations easier, risk assessment robust and investment decisions smarter in the ever-dynamic business environment (Bahrammirzaee 2010). With assistance from machine-learning forecasting tools, companies can produce more accurate forecasts, and allocation of capital resources is done in a more efficient manner as well as precision (Cao 2022). Financial risk management is perhaps one of the most crucial applications of AI. The systems created with artificial intelligence have the ability to search through high volumes of information in order to locate trends that indicate a possible fraud, which is why they improve the security-related procedures and reduce the financial vulnerability significantly (Fritz-Morgenthal 2022). At the same time, AI technology has been viewed to supplement client interaction by providing financial institutions with bottomless client connection and counselling, thus raising customer service and loyalty (Deshmukh et al. 2024). However, the implementation of AI in the financial system is not devoid of major challenges. The question of data privacy, ethical transparency, algorithmic bias and AI interpretability are some key points that limit the ethical implementation of WT. Such issues become especially critical when analysed through the lens of Industry 5.0 the vision of industry in the future that on the one hand it is more human-centred, resilient and ethical to the common benefit (Polak et al. 2020). The current research examines the way that AI is going to develop as a financial management tool in the framework of Industry 5.0. It will first question the aspects of the positive and the negative implications related to the adoption of AI and then will judge whether the benefits exceed the harms, in terms of empirical evidence and contemporary academic literature. This discussion ends with a suggestion of working out the ethical framework of AI adoption where human beings cooperate with AI in such a way that the available financial resources in the economy serve to augment but not to replace, human decision-making abilities in the financial domain. Although much has been done to discuss how AI can be integrated into financial systems, the number of studies that have focused on this is in relation to the Industry 5.0 paradigm. Further, the extent of dearth of empirical studies that concentrate on AI adoption in the Indian financial services market is scarce, as it is a fast growing market that is marked by high rates of technological growth and sophisticated regulatory conditions. The research, in a bid to fill these gaps, provides an integrated theoretical overview and evidence-based research, thus providing a contextualised meaning of the application of human-centric AI in the field of finance.

2.2 Review of Related Literature

Financial management has been transformed by artificial intelligence (AI), and the level of transformation has changed the concept of management of organisations. The digital solutions based on AI are effective systems that can assist in improving operational efficiency, security and accuracy of making decisions. Recent academic literature is about the advantages of the AI application to detecting fraud, automating various parts of the workflow, predictive analytics and improved financial controls; it is the development of Industry 5.0 and creates a new baseline of human-centred and resilient operations. Fraud detection is another important AI usage in finance especially in the case of digital transactions.

A model of online payment fraud detection, created by Almazroi and Ayub (2023), combines Synthetic Minority Oversampling Technique (SMOTE) with ResNeXt and uses gated recurrent units (GRUs), embedded in the first. The research shows that AI-based fraud detection systems can detect suspicious transactions in real-time, minimise rate of false-positives and enhance better security results. Nevertheless, the study is highly performance-focused, and it does not cover the topics of ethical governance, interpretability or human-centric implementation models, which are at the centre of responsible AI integration in finance. Artificial intelligence also promises in the field of financial risk assessment.

Gu (2023) developed an optimised back-propagation neural network (ORP-BNN) to improve the prediction of corporate risks. The results show that machine-learning solutions can enhance accuracy of predicting risk, make the allocation of the funds more efficient and help to make financial practices more robust. This is in line with the model of creating resilience via predictive analytics, which we consider as predictive resilience in the Industry 5.0 model. However, the implications of the model in regard to spirituality of work into organisational structures, regulatory adherence and human-in-the-loop decision-making in financial organisations have not been examined. AI technologies also support decision-making in finance on an enterprise-level. Zhou (2023) examined hybrid fuzzy neural networks, also known as systems that integrate many decision-making methods, and business intelligence platforms as an approach to handling complex financial data.

The findings indicate that AI systems can minimise the number of mistakes and can speed up the decision-making process. In spite of increased speed and accuracy in the study, the areas of explainability, adaptation of the employees and ethical considerations that are vital in sustainable digital transformation have not been addressed. On the other hand, Zhou (2023) investigated the problem of using AI-based clustering methods to facilitate the financial ecosystem of e-commerce across borders. Such methods can improve the level of security of transactions and capital management that are needed in the next generation of globalised financial tools. The platform-based orientation of the paper does not realistic and practical challenges of organisational adaptation, transnational regulation and human-based issues related to the implementation of AI within financial ecosystems. Intelligent support systems can also be used to support strategic financial decisions.

Tong and Tian (2023) examined the impact of big-data analytics and AI on financial planning and shows that the technologies can help in interpreting structured and unstructured financial data to deliver more rational financial information. Simultaneously, the rise of robo-advisory services promotes as one of the AI-enabled financial innovations. Although studies are drawn on the analytical ability of AI systems, they do not allow much attention to how stakeholders interpret, trust or manage such outputs in practice. A study by Ashrafi (2023) investigated the factors that affect the use of AI-robo-advisors and established that consumer trust, financial literacy and risk perception play a critical role in determining the adoption. The paper stresses the fact that honest and transparent AI design can help to increase client trust. Nevertheless, it lacks the institutional governance and system of ethical systems and the transparent adoption of robo-advisory platforms at the organisational level. All these developments have some remarkable advantages but still there exist significant challenges. Zhou (2023) noted that the threat posed by data breaches is increasing rapidly and suggested that AI-based financial systems should have strong cybersecurity. Furthermore, Gu (2023) has highlighted the significance of the model interpretability as the stakeholders should be capable of relying on and interpreting AI outputs. To sum up, the reviewed literature confirms that AI enhances fraud detection, risk assessment and financial decision-making.

However, still, such persistent problems as data privacy, explainability of models, regulatory issues and others are not properly addressed. The given challenges have to be resolved to guarantee the fact that the functioning of AI systems should not only be correct but also moral and concurring to the values of the human society thus fulfilling the goals of Industry 5.0. Although the previous studies have broadly addressed the issue of AI effusiveness and consumers adoption, there remains a significantly large gap in the unification of human-centred and ethical principles based on Industry 5.0. Furthermore, empirical studies situated in the emerging markets specifically in India are old-fashioned, and more so is the research that pulls together the theoretical underpinnings and practice of the institutional practice using the data-driven frameworks. To overcome these limitations, our study will integrate conceptual innovation with empirical evidence in terms of primary concerns to discuss the ethical, organisational and contextual aspects of AI implementation in the Indian financial industry.

2.3 Objectives of the Study

The main aim of the paper is to study how artificial intelligence (AI) may be incorporated into the financial management process, especially how AI-based technologies can help streamline the decision-making process, increase operational efficiency and reduce financial risk. Considering the changing Industry 5.0 paradigm, the purpose of the study is to evaluate how AI may be implemented in a manner that does not only contribute to performance enhancement but also offers the ethical responsibility and human-centric design. To be more exact, this study aims to:

- Examine the empirical cases of AI use in major financial activities such as risk analysis, fraud detection and strategic forecasting.
- Determine the obstacles to the AI adoption in financial organisations, including technological barriers, ethical issues, regulatory uncertainty and staff preparedness.
- Research the effects of AI on jobs in the finance sector and how automation affects the job roles, skills requirements and strategic approaches.
- Suggest a conceptualisation of responsible and transparent AI use, with a focus on explainability, data protection and compliance with legal and ethical requirements.
- Provide value to the discussion about digital transformation by connecting the technological innovation in the financial sector to the human-focused values of Industry 5.0.

2.4 Research Methodology

This paper utilises the descriptive research design, which is based on primary and secondary data. It is proposed to address the ethical, operational or strategic aspects of the artificial intelligence integration in the financial management, especially given the Industry 5.0 principles of transformation that is human-centred.

2.4.1 Primary Data Collection

The primary data for this study was collected through an online structured survey to gather details about how finance professionals use AI tools in their work, their views about how well these tools are working and how they perceive the ethical implications of using AI in their workflow, and what their expectations are about AI tools in their financial work. The survey questionnaire included multiple-choice questions (MCQs) and Likert-scale questions, and it was distributed through a variety of online platforms, including but not limited to email and professional networks.

In total, there were 102 valid responses gathered through non-random sampling. Even though the survey contained sections that asked about the respondents' job titles, industry and level of experience, the focus of this study is on the respondents' interactions with AI tools and their views of the AI tools used by them in their workplace. The survey questionnaire is presented in the appendix.

2.4.2 Sampling and Inclusion

Respondents were selected using a combination of convenience and purposive sampling methods to ensure that participants had sufficient direct experience with financial systems and/or technology systems utilising artificial intelligence (AI). All respondents were included based upon their willingness to participate and their self-identification of relevant experience working in a finance-related environment.

Formal exclusion criteria were not established for this study, although all incomplete responses were eliminated in preprocessing. The respondents included both practicing professionals and graduate/post-graduate students who are engaged in various areas of finance such as banking, investing, consulting and Fintech, with different amounts of work experience. This diversity provides an indication of practitioner views along with the evolving view of emerging professionals.

Due to the fact that this study utilised a non-random, purposive sampling technique, there is a potential for self-selection bias as the participants were recruited via professional networks and volunteerism. Although the sample size of $n = 102$ appears relatively small compared to other studies examining AI adoption in financial management, it matches previous studies examining similar issues in the Indian market where empirical research has historically been limited. Previous studies have also utilised sample sizes of similar size to obtain sector-specific insights. The results of this study are valuable in providing indicative trends; however, due to the non-random nature of the sampling methodology, they should not be considered representative of all possible populations, and subsequent studies should attempt to utilise larger, more diverse samples to increase the level of representation.

2.4.3 Analytical Approach

The statistical analysis of the collected data was organised primarily into descriptive statistics, the basis of which is the determination of frequencies and the derivation of percentage distributions. This method allowed determining general trends related to artificial intelligence adoption, related training activities, user satisfaction and ethical or employment-related issues. These comparative analyses, e.g., comparing perceived effectiveness with increasing levels of exposure to training, were also made.

Moreover, the inferential methods of statistics were utilised to examine the possible interrelations between important variables. The relationships between training status, frequency of the AI usage, level of satisfaction and perceived tool efficacy were assessed by the use of cross-tabulations and 02 tests. Rank correlation was a type of Spearman correlation that was used to investigate the relationship between the ratings of adequacy and effectiveness of training. The calculations were made using inbuilt features in Microsoft Excel.

2.4.4 Secondary Data

Besides the main survey, the research involves the secondary data as the results of scholarly publications, white papers and industry reports. These sources were carefully examined to develop the conceptual framework and to place the empirical results in the context of the broader discourses of AI governance, financial automation and those demands currently emerging in Industry 5.0.

2.5 Conceptual Framework

To develop a comprehensive knowledge of artificial intelligence (AI) in financial management, it is necessary to identify the guiding principles, elements and interrelation structures that define the use of AI in this sector. With financial management being turned into digital and, consequently, aligned with the principles of Industry 5.0, the conceptualisation of AI in financial management should not only extend beyond the purely technical efficiency concerns but also refer to ethical and human factors. The use of data is the heart of this framework since machines can use past and current datasets when identifying trends, assessing risks and providing strategic information to use in financial planning. Deloitte (2022) also notes that the successful use of data in financial planning has the potential to improve investment planning and the overall results on risk management.

AI algorithms and models establish the analytical layer of the conceptual framework. These models are used to process financial data, identify trends and provide insights for strategic financial planning. PwC 2021, reported that more granular customer segmentation (detailed grouping of customers) along with better-targeted financial services through models gives a competitive edge to the financial institutions. An integration component forms the core of the conceptual framework as it represents the integration of artificial intelligence tools into existing financial systems. This involves the modification of workflows, ensuring interoperability between systems and aligning the use of artificial intelligence with the broader financial institution's strategic objectives. McKinsey (2020), reported that integration is a crucial step to ensure that artificial intelligence achieves its full potential in financial institutions.

Ethical considerations are equally important when developing concepts for AI in finance. As more decisions are taken by algorithms, questions around privacy, bias and accountability are emerging. Forum W. E., Artificial intelligence in financial services: Addressing bias, transparency, and data privacy (2023) emphasises the need for governance in order to ensure fairness, transparency and accountability of AI systems and their utilisation in finance. Finally, outcome measurement relates to the ability to measure the tangible results of AI on financial performance.

This could include the degree to which AI leads to efficiency gains, cost savings, improved customer and client satisfaction and improved ability to stay strategic. According to Accenture 2021, successful AI adoption can be seen through the returns achieved by organisations, by better decision-making and higher operational efficiency (Fig. 2.1).

2.6 Data Analysis, Interpretation and Findings

2.6.1 Analysis of Data Collected through Primary Research (N = 102)

These analyses and interpretations are made based on a non-random sample of respondents who completed an online questionnaire that was distributed via email (online circulating) and investigated several areas of inquiry related to the use of artificial intelligence (AI) for the purpose of managing finances, which included but were limited to how frequently AI tools are used, what they are used for and how effective the respondents perceived them to be. In addition to the primary data collected by the respondents of the survey, this section incorporates secondary data from credible and current sources, including both academic and industry research, to provide greater validity and contextually relevant supporting evidence for each identified trend.

1. **Please indicate which AI tools you have used (select all that apply)**: (Table 2.1)

Empirical data shows that the most common artificial intelligence (AI) applications are fraud detection systems followed by chatbots and predictive-analytics applications in financial management. This allocation highlights a strong tendency to use AI in security and customer-engagement capabilities. In turn, automated reporting systems seem to be implemented relatively rare, an indicator of a possible vacuum in the area of financial data automation through AI. The pattern of tool usage follows the norms in the industry. According to the report provided by Bain and Company (2024), 109 financial companies have reported significant productivity improvements because of the implementation of AI in fraud detection and customer-service processes, thus, proving the leading status of these applications in the financial industry.

2. **How frequently do you use AI tools in your work?** (Table 2.2)

A significant number of the respondents indicate occasional or regular use of AI tools, thus demonstrating a two-way split in the rate of AI adoption. Whereas some organisations may integrate AI into normal business operations, some do

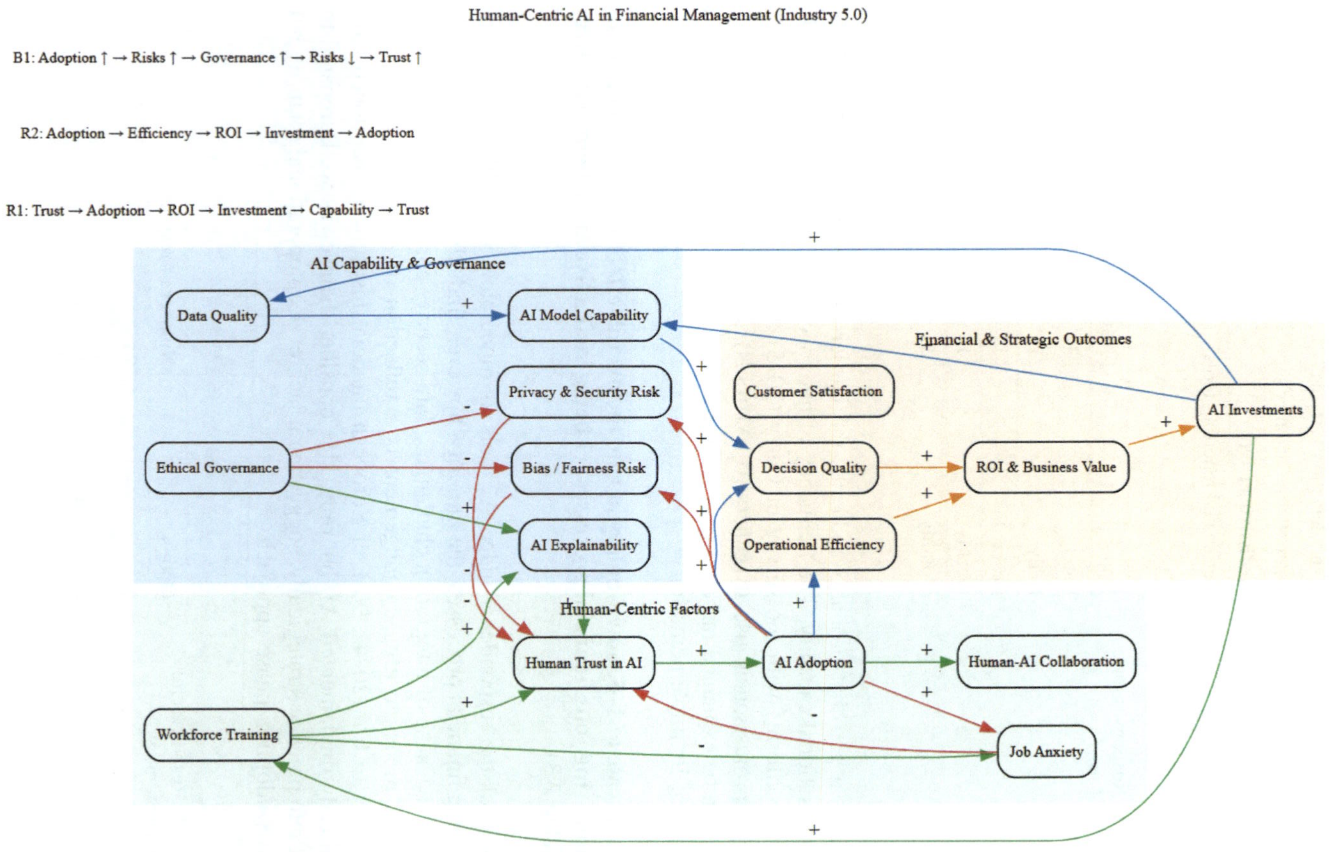

Fig. 2.1 Diagrammatic representation of the conceptual framework using a causal loop diagram: (Author's contribution)

Table 2.1 Responses for AI tools used

Option	Number of responses
Fraud detection systems	34
Chatbots for customer service	25
Predictive analytics	24
Automated reporting tools	19

Table 2.2 Responses for frequency of use

Option	Number of responses
Rarely	32
Daily	25
Never	17
Weekly	15
Monthly	13

it quietly, a gap that probably indicates the barrier of lack of training and lack of organisational capacity. The frequency of heterogeneous utilisation seen in this enquiry conforms with the survey of Curry (2025), which indicates that 52% of financial professionals use it (generative AI tools), but daily integration is rare, which can be explained largely by the infrastructural limitations and the immature maturity of its implementation.

3. **On a scale of 1–5, how would you rate the overall effectiveness of AI tools in improving your financial processes? (1 being ineffective and 5 being effective)** (Table 2.3)

Most of the respondents judge AI tools as moderately or highly effective in improving financial processes; the minority, however, finds them ineffective, which might be because of integration challenges, data-security concerns or lack of training. This moderate efficacy judgement is a reflection of sector-wide reticence. According to EY (2024), 90% of European financial institutions mentioned having already implemented AI, but only 11% consider themselves well prepared to put it into practice strategically, suggesting a lack of deep operational maturity but an early-adoption-phase approach.

Table 2.3 Responses for effectiveness rating

Option	Number of responses
1	24
2	22
3	20
4	18
5	18

Table 2.4 Responses for data-security concerns

Option	Number of responses
Somewhat concerned	28
Not very concerned	21
Neutral	20
Very concerned	17
Not concerned at all	16

Table 2.5 Responses for training received

Option	Number of responses
No	56
Yes	46

4. **How concerned are you about data security and privacy when using AI tools in financial management?** (Table 2.4)

Apprehension regarding data security remains a current concern where most respondents indicated that they are at least partially worried. In spite of the fact that a subset is neutral or unconcerned, there are indications that the organisations should curb security threats in order to create more opportunities of using AI to financial management. These biases in terms of data privacy agree with the World Economic Forum (2025) that presents growing ethical and regulatory risks like opaque algorithms, the possibility of misusing customer data and the lack of explicit governance structure in financial AI systems.

5. **Have you received any formal training in using AI tools for financial management?** (Table 2.5)

More than 50% of the participants indicate having no formal training in AI tools in financial management, which may explain the reported changes in AI implementation and effectiveness. Enhancements of training opportunities could thus reinforce AI use across the industry. The lack of formal training is supported by EY (2024), who stated that a still small proportion of European financial institutions have implemented sufficient internal upskilling programmes to underwrite sustained AI adoption.

6. **How adequate do you find the training and support provided for using AI tools in your organisation?** (Table 2.6)

The current state of artificial intelligence training is a matter of debate, and almost equal numbers of respondents through training rated it as very adequate or very inadequate. Such dispersion indicates variations in the AI training programmes in organisations, and, therefore, standardised AI training is important. The polarised answers about the adequacy of training reflect the fragmentation on

Table 2.6 Responses for training adequacy

Option	Number of responses
Very adequate	22
Very inadequate	22
Neutral	21
Somewhat adequate	19
Somewhat inadequate	18

Table 2.7 Responses for impact on job role

Option	Number of responses
Increased efficiency	33
Job displacement	24
More strategic focus	23
No significant impact	22

the study of SME Economic Forum (2023), ranking inconsistent training sources and cost-linked obstacles to workforce preparedness to integrate AI as significant setbacks.

7. **How has AI impacted your job role in financial management?** (Table 2.7)

AI has had a diverse influence on the financial management functions, and the most commonly mentioned benefit is efficiency. However, the job displacement issues remain, which means that organisations have to strike the balance between AI-based automation and job adaptation plans. The fear of job loss indicated in the data of the survey corresponds to the forecast of Citi (2024) (according to which over 50% of all banking jobs globally stand a high probability of being replaced by computers), which in turn has to denote that the reconsideration of human and computer cooperation in financial institutions should be taken seriously.

8. **Overall, how satisfied are you with the AI tools currently available for financial management?** (Table 2.8)

Table 2.8 Responses for satisfaction with tools

Option	Number of responses
Very dissatisfied	22
Dissatisfied	22
Neutral	22
Very satisfied	19
Satisfied	17

Table 2.9 Responses for future growth of AI

Option	Number of responses
Yes	39
Unsure	38
No	25

The attitudes towards AI tools vary as just over half of all surveyed expressed satisfaction, and an equally high number were neutral. This trend indicates that, although AI can have a measurable advantage, there is still so much room to enhance functionality, convenience and consistency. The discrepancy in levels of satisfaction is supported by PYMNTS (2024), which has found that banks that describe their levels of AI satisfaction as less have problems with the integration process, undefined AI policy and untrained staff that reduces the ROI and confidence of employees.

9. **Do you believe that the role of AI in financial management will grow in the next 5 years?** (Table 2.9)

Most of the respondents assume that AI will become more and more prominent in financial management within five years, but a considerable fraction of participants are not sure. This underscores the importance of more studies and better directions in the industry so that the implementation of AI is compatible with business goals. Reuters (2023) noted that the broadly optimistic future view of AI in finance is in line with JPMorgan (2025) revealing that AI systems were vital in the client and strategic decision-making in turbulent market conditions. Namely, the Coach AI by JPMorgan allowed advisers to find research and anticipate investment questionings up to 95% faster, which was directly related to client trust and responsiveness. It has helped this company to grow its gross sales in its asset and wealth-management department by 20% between 2023 and 2024. Already with a technology budget of more than 17 billion dollars and more than 450 known AI uses, the company has already saved the company almost 1.5 billion dollars due to applications in fraud prevention, personalisation, trading efficiency and credit decisioning. These results demonstrate that AI implementation in the financial field is both empowering and disheartening, as it enhances development and fortification on the one hand and requires a greater number of staff training, compliance with regulatory regulations and a solid ethical leadership on the other hand, which resonates with our survey results (Table 2.10).

2.6.2 Inferential Analysis

To go beyond descriptive trends, inferential tests were conducted to identify statistically significant relationships between variables in the survey (Table 2.11).

Table 2.10 Comparative AI adoption trends across geographies and sub-sectors

Region/sub-sector	Adoption focus areas	Adoption level/key findings	Sources
India (Survey, 2025)	Fraud detection, chatbots, predictive analytics	Uneven adoption levels: 25% daily versus 32% rarely; significant training gap	Present study
Europe (Financial Firms 2024)	Risk modelling, compliance automation, GenAI pilots	90% adoption, but only 11% strategically mature	EY (2024)
United States (retail banking)	Client engagement, fraud prevention, personalisation	20% gross sales boost (2023–24); ~ $1.5B savings; 200 k staff AI toolkit rollout	Reuters (2025)
Global investment firms	Algorithmic trading, risk prediction, portfolio optimisation	High adoption, but challenges in explainability, governance, regulatory compliance	WEF (2023, 2025)
SMEs in finance (global)	Basic automation, entry-level chatbots	Cost and training barriers; fragmented adoption patterns	SME Finance Forum (2024)

Table 2.11 AI training × effectiveness rating

Effectiveness rating	No training (n = 56)	Training (n = 46)	Total
1—Not effective	10	2	12
2	12	6	18
3	15	3	18
4	9	11	20
5—Very effective	10	12	22
Total	56	46	102

A Chi-square test indicated a statistically significant association between training and perceived effectiveness, $\chi^2(4, N = 102) = 12.76$, $p < 0.05$. Respondents with training had a higher average effectiveness rating (4.1) compared to those without training (3.4).

Training Adequacy × Effectiveness Rating

Spearman's rank correlation between training adequacy (1 = very inadequate to 5 = very adequate) and effectiveness rating (1–5 scale) was $\rho = 0.48$, $p < 0.001$, indicating a moderate positive relationship. Higher training adequacy was linked to higher perceived tool effectiveness (Table 2.12).

Table 2.12 Frequency of AI use × satisfaction with AI tools

Satisfaction level	Rarely	Monthly	Weekly	Daily	Never	Total
Very dissatisfied	8	2	3	2	7	22
Dissatisfied	9	3	3	2	5	22
Neutral	8	4	4	3	3	22
Satisfied	3	3	3	4	4	17
Very satisfied	4	1	2	14	0	19
Total	32	13	15	25	19	102

The Chi-square result was $\chi^2(8, N = 102) = 15.21$, $p \approx 0.055$. While just above the conventional significance threshold, a clear trend emerged: daily AI users were more likely to report high satisfaction.

Interpretation
The results of the inferential analysis put forward the fact that AI training and adequacy of training are majorly linked to the higher perceived effectiveness, while a frequent use of artificial intelligence is linked to a greater degree of satisfaction. This supports the suggestion for including not only targeted upskilling programmes but also consistent AI engagement within the AI institutions.

2.6.3 Findings

Leadership in Security and Advisory Application: According to the existing report, applications that are customer and security focused are one of the most common utilised artificial intelligence (AI) applications in the financial sector. The most mentioned AI technology is the one dealing with fraud detection, which is reported by about one out of three of all respondents. Next, chatbots and predictive analytics are mentioned. These results support the notion of bolstering security and improving the customer experience, as well as indicate that there is a relative lack of AI use in back-office activities like automated reporting, thus indicating implementation gap in financial operations.

Polarised Pattern of Use: The report states that the use of AI in the sector is polarised. Although one in every four financial professionals report using AI on daily basis, one-third of the respondents report having to use AI rarely. Such a difference implies the presence of a large gap between the digital maturities of organisations. Besides, the report refers to infrastructure and organisational limitations that distinguish between firms that actively use AI technologies and others.

Moderate and Majorly Positive Views of Tool Effectiveness: In general, the financial professionals assess their AI tools as fairly to very effective as regards to financial management. However, a good percentage of the participants are unhappy with AI tools with references to its inefficient application, inaccurate data and insufficient training. These problems are a note to the fact that the introduction of AI into a financial system does not necessarily imply a full-fledged continuum of the digital environment and a corresponding improvement in performance.

Fear of Data Privacy and Algorithms Transparency: According to the survey results, there is a substantial fear of data privacy and algorithmic transparency when it comes to utilising AI tools, which is due to predictable results. About 45% of the participants are concerned with the threat of data security. These concerns represent international concerns about the absence of transparency in algorithms, prejudice and the possibility of non-observation of the standards of data protection. These issues highlight the necessity to initiate new approaches to implement ethical AI practice, particularly in finance with high stakes.

Another Important Trend of Training to Use AI: The data collected through the survey shows a strong training gap, as more than half of the respondents story about a lack of formal training in AI. This gap is an obstacle to the proper implementation of AI products and, consequently, to the productivity and trust of consumers. The participants of the survey support the use of training programmes aimed at educating financial professionals to take the best advantage of their AI tools.

A Divided Response on Adequacy of Training on AI-Related Topics in Organisations: In responding to whether the training they received on AI-related topics in their organisations was adequate or not, the response was almost 50 to 50 in between answers of very adequate and very inadequate. The findings underscore the need for a standardized AI training framework to ensure that employees are adequately prepared to work with AI technologies.

Creative Complicatedness in Job Objective Because of the Impact of AI: The work in the financial sector does not seem easy. One-third of the respondents state that AI has made their operations more efficient, and 24% of people believe that their jobs are under risk of being replaced by AI. Another 23% report that their work has been more strategic, as routine work is taken over by AI applications. Taken together, these results demonstrate the two-sided character of AI in its impact on the job role: on the one hand, the technology can increase its efficiency, and on the other, it can jeopardise the jobs and require retraining.

Partial Satisfaction About AI Tools: The respondents admit that AI tools are useful and needed; nevertheless, the general satisfaction is low. One-third of the respondents claim to be unsatisfied or deeply unsatisfied with the tools at his disposal. This feeling poses a possibility of enhancing the user interface design, ease of use and customisation of tools to be more in line with the needs of financial professionals.

Opinion About the Future Role of AI in the Next Five Years: The future growth of AI in financial management is identified by a good percentage of the respondents (39%). However, a similar percentage expect a reduction of the role of AI, and almost the same number expect to remain unsure of the future implications of AI in financial management (38–39%). Though there is optimism on the future of AI in the industry, strategic focus is necessary to make AI fulfil its widened role.

2.7 Conclusion

There is a paradigm shift in the financial industry, and the key driver of change is the artificial intelligence (AI). The practice shows that AI is no longer speculative; it can be actually used to benefit financial institutions through detecting frauds, forecasting risks, aiding decision-making and serving customers. However, the implementation of AI in the industry is uneven, and there are many challenges disabling the process of implementation. As it can be seen, AI increases the data precision, shortens the processing time and advances the understanding of data. Nonetheless, these quantitative advantages cannot be achieved through technological infrastructure alone. Successful AI implementation also requires adequate organisational support, employee readiness, and effective change management. The basic capital is human capital. Most of the people working in the industry have had some training, but a large number are inexperienced.

Difficulties in data confidentiality and displacement of jobs have continued to be of concern. Both technical and human and societal issues surrounding ethical considerations of algorithmic decision-making are not only technical issues but deeply human ones that demand transparency and trust. Complete automatisation and profit maximisation are not the sole goals of the future AI in finance. Rather, the focus should be to align these goals with the greater benefits of society. Industry 5.0 principles can be used as a guideline since the meaning of technology, according to them, is to promote societal and social development instead of acting to speed up the machinery. The AI will make an improved system faster and more accurate, but it will also be influenced by the experience, preferences and intuitions of financial experts. The sector will have to adjust to this change by training employees and setting up effective data management regulations and making AI decision-making a transparent process.

The model of the AI as an additional tool to human strength enables finance practitioners to shift their mindsets into strategic and collaborative thinking instead of focusing on the routine workload. Ethical grounds of AI should be given beforehand, thus forcing decision-makers, regulators and educators to consider not only what AI does but also how it operates to ensure it keeps abreast of technological progress.

2.8 Recommendations

Popularise Structured AI Literacy in Financial Institutions: Financial institutions are supposed to ensure that an individual in all hierarchies has a solid base of knowledge on the concepts of AI. It is proposed to integrate in-depth learning modules into the training programmes in organisations where a considerable percentage of the staff do not have any prior formal exposure to AI. Such research programmes must include technical and non-technical staff to facilitate human judgement in complementing machine outputs.

Establish Institution-Specific Ethical Governance Commissions: To mitigate the risks of having opaque, biased and misused data, the establishment of generic compliance is insufficient. Both the institutions must develop context-specific ethical principles that guide the design, adoption and management of AI. Transparency, accountability, explainability and fairness should be established by means of strict internal checks and external audits.

Restructure Organisational Design to Improve Human-AI Cooperation: Instead of viewing AI adoption only as a cost coding procedure, organisations ought to value its ability to improve the performance of the human staff. This necessitates re-evaluation of organisational attitudes that have traditionally led to automation-inspired restructuring. The organisation of job roles must be redesigned to require human judgement, analysis and long-term projection, and the routine processes may be assigned to AI under human control.

Add Feedback Loops and Outcome Monitoring Mechanisms to Organisations: AI implementation is not a single event but a process that advances repeatedly. The organisations ought to establish mechanisms to assess AI tools on a regular basis. These mechanisms are expected to extend across the wider institution and society, addressing implications on employee welfare, customer experience and ethical adherence. Feedback of internal and external stakeholders should be regular and repeated to shape the constant improvement of AI-related strategies.

Develop Data Governance Policies and Invest in Cybersecurity: Due to the increasing worries surrounding data protection, organisations need to institutionalise the centralised data governance policies, which would cover data ownership, consent, access management and anonymisation among other privacy concerns. Cybersecurity tools should be invested in concurrently when it comes to decreasing vulnerabilities associated with real-time AI systems.

Promote Interdisciplinary Cooperation in the Design and Supervision of AI: AI in finance should not be left to data scientists or IT specialists alone. The potential ethics behind the use of AI applications require and promote interaction between legal experts, behavioural scientists and ethicists, seeing to it that AI solutions are not merely technologically valid and sound, but also socially accountable and attentive regarding dynamic stakeholder needs.

Promote Sector-Wide Discussion and Policy Coherence: To prevent the challenges of siloed AI strategies and regulatory uncertainties, there must be coordinated industry and country efforts. Financial regulators, academia and industry associations must all work together to identify common standards, open-access research and strong knowledge transfer frameworks.

Table 2.13 Key recommendations summarised

Key finding	Implication	Recommendation
Polarised AI adoption (daily vs. rare users)	Uneven digital maturity	Sector-specific AI adoption roadmaps
Lack of training (56/102 untrained)	Lower effectiveness and satisfaction	Standardised AI training/ upskilling programmes
Data privacy concerns (45% concerned)	Trust deficit	Establish AI governance and ethical oversight bodies
Mixed satisfaction with tools	Tool immaturity and poor integration	Vendor-institution co-design for usability and explainability
Job displacement fears	Workforce anxiety	Reskilling initiatives + human-AI collaboration models

Research Instructions Must Be Contextual and Longitudinal and Must Be About Humans: Although the current research offers valuable technical information, future studies need to examine the human impact of AI on organisational culture, work and ethics decision-making in the long term. Comparisons between financial sub-sectors and geographies would help clarify the effects of institutional context on AI. Moreover, client attitude towards trust, inclusion and financial well-being would be interesting to consider to enhance the discussion. Longitudinal studies aimed at assessing the results of the proposed collaboration between humans and AI, including trust metrics (e.g. perceived transparency, fairness, explainability), productivity metrics (e.g. time to complete tasks, error reduction, strategic decision quality) and indices of employee adaptability, would provide broad evidence regarding the development of a new synergy between human professionals and AI systems, going beyond the short-term efficiency benefits (Table 2.13).

Appendix—Questionnaire

Personal Information

(i) Name: ______________________________.
(ii) Job title: ___________________________.
(iii) Company: ____________________________.
(iv) Industry: ____________________________.
(v) Years of experience in finance: ________.

Q. Please indicate which AI tools you have used (select all that apply)

- Predictive analytics.
- Automated reporting tools.

- Fraud detection systems.
- Chatbots for customer service.
- Others: ________________.

Q. How frequently do you use AI tools in your work?

(a) Daily.
(b) Weekly.
(c) Monthly.
(d) Rarely.
(e) Never.

Q. What benefits do you associate with using AI in financial management? (select all that apply)

- Improved accuracy in data analysis.
- Time savings in reporting.
- Enhanced decision-making capabilities.
- Reduced operational costs.
- Better risk management.
- Others: ________________.

Q. On a scale of 1 to 5, how would you rate the overall effectiveness of AI tools in improving your financial processes?

(a) 1 (Not effective)
(b) 2
(c) 3
(d) 4
(e) 5 (Very effective).

Q. What challenges have you faced in implementing AI in your financial practices? (Select all that apply)

(a) Lack of technical expertise.
(b) Resistance to change within the organisation.
(c) High costs of implementation.
(d) Data privacy concerns.
(e) Integration with existing systems.
(f) Others: ________________.

Q. How concerned are you about data security and privacy when using AI tools in financial management?

(a) Very concerned.
(b) Somewhat concerned.
(c) Neutral.
(d) Not very concerned.
(e) Not concerned at all.

Q. Have you received any formal training in using AI tools for financial management?

(a) Yes.
(b) No.

Q. How adequate do you find the training and support provided for using AI tools in your organisation?

(a) Very adequate.
(b) Somewhat adequate.
(c) Neutral.
(d) Somewhat inadequate.
(e) Very inadequate.

Q. What has been the outcome of integrating AI into your financial management job?

(a) Increased efficiency.
(b) More strategic focus.
(c) Job displacement.
(d) No significant impact.
(e) Others: ________________.

Q. What is your level of satisfaction with the AI tools currently available for financial management?

(a) Very satisfied.
(b) Satisfied.
(c) Neutral.
(d) Dissatisfied.
(e) Very dissatisfied.

Q. Do you think that the role of artificial intelligence will expand in the coming 5 years?

(a) Yes.
(b) No.
(c) Unsure.

Q. Do you have any suggestions or recommendations concerning the improvement of AI tools in financial management?

Q. Please share any personal insights on AI's future in the finance sector!

References

Accenture (2021) The art of AI maturity: advancing from practice to performance. Accenture Insights. Retrieved from https://www.accenture.com/us-en/insights/artificial-intelligence/ai-maturity-and-transformation

Almazroi AA, Ayub N (2023) Online payment fraud detection model using machine learning techniques. IEEE Access. https://doi.org/10.1109/ACCESS.2023.3339226

Ashrafi DM (2023) Managing consumers' adoption of artificial intelligence-based financial robo-advisory services: a moderated mediation model. J Indonesian Econ Bus 38(3):270–301

Bahrammirzaee A (2010) A comparative survey of artificial intelligence applications in finance: artificial neural networks, expert systems, and hybrid intelligent systems. Neural Comput & Applic 19(8):1165–1195. https://doi.org/10.1007/s00521-010-0362-z

Bain & Company (2024) AI in financial services survey shows productivity gains across the board. https://www.bain.com/insights/ai-in-financial-services-survey-shows-productivity-gains-across-the-board/

Cao L (2022) AI in finance: challenges, techniques, and opportunities. ACM Comput Surv 55(3):1–38. https://doi.org/10.1145/3502289

Curry R (2025 Feb 13). NVIDIA exec on what's next for AI adoption in financial services. Innovation Leader. https://www.innovationleader.com/topics/articles-and-content-by-topic/scouting-trends-and-tech/nvidia-on-ai-in-financial-services/

Deloitte (2022) Managing gen AI risks. Deloitte Insights. Retrieved from https://www2.deloitte.com/us/en/insights/topics/digital-transformation/four-emerging-categories-of-gen-ai-risks.html

Deshmukh PB, Nikam P, Moholkar N (2024) Role of artificial intelligence in financial management. ShodhKosh J Visual Perform Arts 5(6):541–557. https://doi.org/10.29121/shodhkosh.v5.i6.2024.1765

EY (2024, Dec) Lack of upskilling and low regulatory preparedness is holding back deeper GenAI adoption within European financial services firms. https://www.ey.com/en_gl/newsroom/2024/12/lack-of-upskilling-and-low-regulatory-preparedness-is-holding-back-deeper-genai-adoption-within-european-financial-services-firms

Fritz-Morgenthal S, Hein B, Papenbrock J (2022) Financial risk management and explainable, trustworthy, responsible AI. Front Artif Intell 5:779799. https://doi.org/10.3389/frai.2022.779799

Gu L (2023) Optimized backpropagation neural network for risk prediction in corporate financial management. Sci Rep 13:19330. https://doi.org/10.1038/s41598-023-46528-8

McKinsey & Company (2020) Building the AI bank of the future. McKinsey Global Institute. Retrieved from https://www.mckinsey.com/~/media/mckinsey/industries/financial%20services/our%20insights/building%20the%20ai%20bank%20of%20the%20future/building-the-ai-bank-of-the-future.pdf

Polak P, Nelischer C, Guo H, Robertson DC (2020) "Intelligent" finance and treasury management: what we can expect. AI & Soc 35:715–726. https://doi.org/10.1007/s00146-019-00919-6

PwC (2021) Risk and regulatory outlook 2021: use of AI/ML and alternative data in credit decisioning. PwC Financial Services Report. Retrieved from https://www.pwc.com/sg/en/insights/assets/docs/risk-regulatory-outlook-2021-use-ai-ml-alternative-data-credit-decisioning.pdf

PYMNTS (2024, June 19) Citi: AI will impact finance sector job market the most. https://www.pymnts.com/personnel/2024/citi-ai-will-impact-finance-sector-job-market-the-most/

Reuters (2025, May 6) JPMorgan says AI helped boost sales, add clients in market turmoil. https://www.reuters.com/business/finance/jpmorgan-says-ai-helped-boost-sales-add-clients-market-turmoil-2025-05-05/

SME Finance Forum (2024) State of AI in financial services: 2024 trends. https://www.smefinanceforum.org/sites/default/files/post/files/finance-state-of-ai-report-2024-3067247%20(1).pdf

Tong D, Tian G (2023) Intelligent financial decision support system based on big data. J Intell Syst 32(1):20220320. https://doi.org/10.1515/jisys-2022-0320

World Economic Forum (2023) Artificial intelligence in financial services: Addressing bias, transparency, and data privacy. Retrieved from https://reports.weforum.org/docs/WEF_Artificial_Intelligence_in_Financial_Services_2023f

World Economic Forum (2025) Artificial intelligence in financial services. https://reports.weforum.org/docs/WEF_Artificial_Intelligence_in_Financial_Services_2025.pdf

Zhou K (2023) Financial model construction of a cross-border e-commerce platform based on machine learning. Neural Comput & Applic 35:25189–25199. https://doi.org/10.1007/s00521-023-08456-6

From Algorithms to Advisors: Exploring AIs Dual Disruption in Trading and Wealth Management

3

Vaibhav Aggarwal, Mahender Yadav, Shashank Sharma, and Reepu

Abstract

Over the past decade, it has become evident that the financial sector's landscape has undergone significant changes due to the advent of Artificial Intelligence (AI) and Machine Learning (ML). These twin technologies are responsible for creating a disruptive technological advancement in both the trading and wealth management verticals via algorithmic trading and Robo-Advisors, respectively. The usage of these technologies is providing several benefits and at the same time opening the doors for regulatory challenges. The benefits and challenges of these technologies are discussed through the lens of financial and technological adoption theories such as Efficient Market Theory, Behavioural Finance Theory, Innovation Diffusion Theory, and Principal-Agent Theory. This chapter of the book is influenced by Jane Street, a big international trading platform that earns billions of profits in the Indian Market in a fraction of seconds, which raises the concerns of Security Exchange Board of India, that such high-frequency Algo-based trading will effect retail investors. This study highlights the benefits and challenges and also provides the pathways to regulators to cope with this dilemma.

V. Aggarwal (✉)
O.P. Jindal Global University, Sonipat, Haryana, India

M. Yadav
Maharshi Dayanand University, Rohtak, India
e-mail: mahinder.comm@mdurohtak.ac.in

S. Sharma
JPMorgan Chase & Co, New York, USA

Reepu
Chandigarh University, Mohali, Punjab, India

B. Kaur et al. (eds.), *Integrating Industry 5.0 and 4.0 for Technology and Humanity*, Contributions to Environmental Sciences & Innovative Business Technology,
https://doi.org/10.1007/978-981-92-1780-9_3

Keywords

Artificial intelligence • Machine learning • Algorithmic trading • Robo-advisory • Challenges

3.1 Introduction

In the Indian Market, the Security Exchange Board of India (SEBI) constantly provides strict guidelines for Algo traders in order to make the market inclusive, transparent, and stable and finally protects the rights of the retail investors (Parashara et al. 2024). The concerns started from the suspicious trading done by one of the Algo trading companies, Jane Street, which made a massive profit of Rs 735 crore in a single day. This study provides the role of Artificial Intelligence (AI) and Machine Learning (ML) in the financial service industry, particularly in the capital market via stock trading and in the financial wealth management industry. This is imperative to understand how these industries have been changing and boost the financial market; however, it also highlights the risks and concerns in both verticals of the financial industries. The study also highlights the immediate concerns of regulators and finally provides the pathways to mitigate the risks of Algo trading (Ramkumar 2018) and Robo-Advisory-based financial planning.

At the outset, it is worth mentioning that AI has changed the financial industry (Sharma et al. 2022), particularly in stock trading and wealth management (Fatima and Chakraborty 2024). In India, Algo trading plays a significant trigger in increasing the trading volumes (Riyazahmed 2024, Dubey et al. 2017) that is accounting for 57% of equity cash and 70% of Future and Option (F&O) trades in mid-2025 (Moneycontrol 2025). This not only accentuates the volume but also increases liquidity, efficiency, and market speed. Moreover, India is catching up to developed markets US and Europe by 60–85%. The growth of AI and ML in financial sector is because of the usage of cloud computing (Marketsandata 2025). However, Algo trading, despite of enhancing market efficiency, it also provide risk and losses to retail investors (Dubey et al. 2022). It has been believed that Algo trading that in India, the algorithmic trading is projected to grow at a Compounded Annual Growth Rate (CAGR) of 11.65% within the forecasted period of FY 2025 to FY 2032. It has been growing from USD 1.08 billion in FY2024 to USD 2.61 billion in FY 2032. The driving factors that triggers its growth are many, firstly, the increased access of high-frequency data through API keys and cloud-based solutions, secondly, rising Algo platforms supported by fintech, third is rising disposable income and lower transaction cost, and the most important trigger is regulators support.

At the same time, AI has captured another prominent sector i.e. wealth management (Kesharwani et al. 2024). In this sector, Robo-Advisors are prominently capturing the advisory assistance, providing instant automated financial plan at a low cost and having a larger reach to customers and thus replacing the traditional wealth management advisors. It's a new edge of investments and thus, popular among millennials and new tech edge investors (Chandani et al. 2021). The platforms are using AI and Machine Learning algorithms for client profiling, understanding their risk appetite, capturing their financial assets and liabilities, and their income and expenses, finally providing the financial plan and creating their sophisticated portfolio management (Nain 2023). The most important aspect is that the technology is providing advisory that is free from human bias. The technology is very popular in developed markets, for instance, in the United States, the Big 4 banks are using Robo-Advisors and providing financial advisory services to their customers at high speed and low cost. However, in India, it's in the nascent stage and capturing the market steadily, on the ground of its high speed and lower cost. In India, the traditional advisors are charging 1–2% or more for Portfolio Management Services (PMS) and active strategy in financial markets and funds that have been reduced significantly (Chandani et al. 2021).

However, the contribution of AI is now of concern in financial markets after the incident of Jane trader who secured massive billions of profits in a single day trade. The dilemma the regulators are facing is whether to embrace the opportunities providing by AI in the financial sector and, in the same hand, securing the interest of retail investors. In the year 2015, the SEBI has provided a comprehensive regulatory framework for Algo trading to protect the interests of investors from unreasonable volatility, probable losses, and ensuring market integrity. This chapter highlights the key benefits, concerns, and finally provides a pathway to mitigate the risk arising and thus, securing the interest of retail investors.

3.2 Theoretical Framework

The benefits and challenges of AI in the financial industry can be explained from various financial and technology adoption theories like Efficient Market Theory, Behavioural Finance Theory, Innovation Diffusion Theory, Principal-Agent Theory, and finally Regulatory Compliance Theory.

1. **Efficient Market Theory (EMH)**—The theory was propounded by Fama (1970). The theory emphasizes that the market are efficient, which means all information is reflected in the prices of assets. As per the theory, investor cannot beat the market (Malkiel and Fama 1970). As per this, technical analysis is a failure strategy to get benefit or profit based on past patterns. It concluded that both technical and fundamental strategies are futile and thus unable to provide any alpha from the market. However, this theory has been challenged by the Algo trading where the traders are making a strategy based on technical indicators and also supports and validate their strategy with back testing. They created predefined rules, and on the basis of set rules of buy and sell, the

Algo works independently without human interventions. The Jane Street incident is best example that rejects the theory of efficient market, which runs on the hypothesis that no one can beat the market or earn abnormal gains.
2. **Behavioural Finance Theory**—The theory was given by Daniel Kahneman, Amos Tversky, Richard Thaler, and Robert Shiller (Illiashenko 2017). The theory suggests that investors do not always act rationally. They are influenced by psychological biases, social influence, emotions, and mental shortcuts, and because of this market becomes inefficient. It leads to mispricing, bubbles, and crashes. The theory has been contradicted by Algo trading and Robo-Advisors that work on automated well-defined formulae and algorithms that are free from human interventions and thus free from human bias and emotions. It makes the market efficient, thus reducing the chances of panic selling, overconfidence, and herd behaviour (Bhatia et al. 2020).
3. **Innovation Diffusion Theory**—The theory was given by Rogers and Cartano (1962), highlights how new ideas, technologies, or new practices spread or are absorbed by the people. The theory says that the speed of adoption varies, and it depends on characteristics, social influence, and perceived benefits. The Algo trading and Robo-Advisors adoption has been triggered by its relatively lowered cost comparative to the traditional system, and another factor could be the compatibility of adoption of AI platforms backed with an efficient trading system and support of regulators. However, it can also show delayed adoption because of its complexity and ethical concerns.
4. **Principal-Agent Theory**—The theory was developed by economists Michael and William. The theory explains the relationship between principal and agent, the principal is known as the owner, and the agent is known as the person who is hired on behalf of the owner. The theory focusses on the conflict in between principal and agent because of their conflicting interest. This theory has been applicable to Algo trading and Robo-Advisors, both are replacing the traditional way of trading that may mislead investors because of their commissions. In the context of trading, the brokers are sometimes making unnecessary trades to get their brokerage charged on each trade; on the other hand, human financial advisors may also do misleading financial plan to get their commissions. In contrast to this, Algo trading and Robo-Advisors will provide more transparency and reduce this conflict.

3.3 Conceptual Framework

The conceptual model (Fig. 3.1) captures how disruptive technologies like AI and ML acted as an enabler to change the landscape of the financial industry. These twin technologies are the basis of both evolutionary sub-technologies, which are Algo trading and Robo-Advisory. Both sets of platforms are worked on using AI technologies and predefined ML-based algorithm models extensively.

Algo trading platforms by leveraging AI and ML technologies can do transactions in fractions of a second. Thus, promoting high speed, increasing its reach,

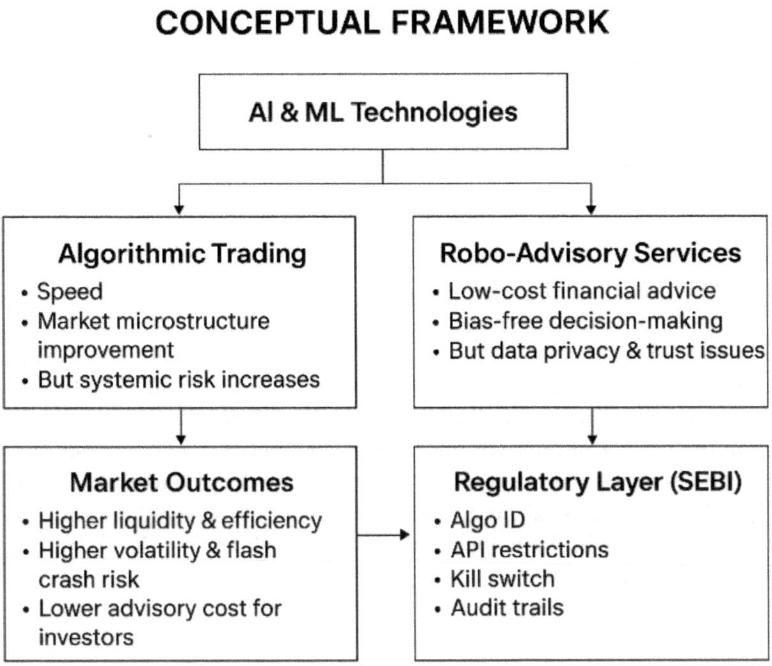

Fig. 3.1 Conceptual framework. *Source* Authors own contribution

and makes the transaction affordable and less costly. On the other hand, it increases the systematic risk to retail investors, whereas, Robo-Advisors by using twin technologies, replace the traditional human led financial advisors, and thus now provide financial advice and prepare their robust portfolios at very low cost. The advice has been given by the machine on a predefined algorithm, thus it is biased free. However, advisors need customers' data such as demographics, economic, and financial data, thus it is exposed of ethical concerns like data privacy.

These two financial AI-driven platforms, particularly Algo trading, provide higher liquidity and efficiency to the market, reducing brokerage and advisory cost, free from bias. But at the same time, it accentuates the volatility that may lead to systematic risk. This triggers the involvement of regulators. Henceforth, SEBI forms the regulatory layer to protect the interests of retail investors. This has been activated more rigorously after the fraudulently participation of Jane Street, an international Algo trading company, booked crores of profit in a single day. Thus, to balance innovation with investor protection, SEBI has taken measures like Algo ID, API restriction, Kill switch, and Audit trails to safeguard the retail investors from any systematic risks emerged from Algo trading.

To conclude, the conceptual framework shows how two disruptive technologies gave birth to two AI-driven innovative platforms that leverage two financial segments, but at the same time provide volatility and introduces risks. In this context, regulators acted as a stabilizer and struck a balance between innovations and stability and market integrity.

3.4 Algorithmic Trading—Mechanism, Benefits, and Risk

Algorithmic trading is based on predefined rules set by the trader based on mathematical models, technical analysis rules, statistical analysis driven rules, and Machine Learning models. The strategies that are made are tested well on the historical data, to make the strategy tried and tested, and then implemented in the current financial market. Instead of a human trader, the buying and selling orders have been done by a machine based on predetermined, well-tested rules. Thus, the algorithm automates the entire process and scans the wide range of securities trading in the market in microseconds, and if the predetermined formula fits in the given security, it automatically runs the trade. The Algo trading mechanism has been based on four crucial components: instructions received by the algorithm, execution of orders with high speed, performing back testing, designing a strategy, and reducing latency.

3.4.1 Algorithmic Trading—Mechanism and Benefits

Algos are working on algorithm that are based on predefined rules, identifying the right timings of the market and then triggering the orders. These rules are created by traders and data scientists that are based on technical indicators like Moving Averages, Candle stick patterns, Relative Strength Index, and other techniques. They are using statistical and time series analysis and other forecasting techniques such as mean reversion models, Autoregressive Moving Averages, Generalized Autoregressive Heteroscedasticity models, and other advanced econometric models. Algo traders are also using Machine Learning Models such as Artificial Neural Networks, Long Short Term Memory, and other supervised and unsupervised techniques to predict the prices and finding the right time to enter into a trade. The Algo trading platforms also signal the arbitrage opportunities by scanning the market and detecting the mispricing opportunities and executing the trade. The platforms are also following risk management techniques like stop loss and also taking a position of hedging. These well set rules have been coded in software like Python, C++, or R, or executed through trading APIs. Once the rules are coded, the algorithm consistently scanning the market, and wherever the machine is finding the rule is applicable and it triggers the orders.

The Algo trading is executing the orders at its extremely high speed in microseconds, called high-frequency trading (HFT). The algorithmic trading executes its orders in nanoseconds to microseconds, and this is only possible because of its ecosystem (Aggarwal et al. 2023; Dhingra et al. 2023; Nahar et al. 2024). It has a high speed computing infrastructure, the servers are located close to the stock exchanges, which reduces its latency, and this reduces its time to execute the orders. Another factor is that it feeds real-time data via APIs; with the help of this technology, the platform is able to work on real-time market data, and every tick is captured on a real-time basis. Moreover, the entire order was executed automatically without human intervention. The order has been executed through flawless routes and done in a fraction of seconds. The order is first generated by the trading engine, has been sent by the broker's server, then reaches to exchange, and finally confirmation is received that the order is executed.

There are several benefits provided by Algo trading. It provides liquidity and efficiency in the market. For instance, the participation of institutional investors, large orders are traded, and applied various risk management techniques. At the same time, it creates volatility in the market. For instance, say $100 million order for a stock is entered to hedge a previous buy order—this will lead to a spike in price and accentuates high impact costs. The Algo platforms assist them in deploying the hedging in various small orders that minimize the market volatility and reduce the impact cost. Henceforth, Algo platforms are used to provide stability in the market, execute orders in a systematic way with relatively lower impact cost, and finally reduce human emotional bias.

3.4.2 Systemic Risk Associated with Algo Trading

The systemic risk of widespread algorithmic adoption is a pertinent question that needs continuous regulatory vigilance and framework. The risk associated with Algo trading are varied, for instance, the interconnectedness and speed of algorithms mean that a sudden market shock can trigger simultaneous buying or selling which can amplify volatility and increase the risk of systemic risks, often culminating in 'flash crashes'. Past studies highlighted that although automation reduces human intervention, it also introduces a new class of systemic risk (Akansu 2017; Kirilenko et al. 2017). Another risk emerges from applying advanced AI, in multi-agent trading environments. AI algorithms can, through self-learning and adaptation, learn to implicitly coordinate or collude on trading strategies, even without explicit instruction or communication. This emergent, coordinated behaviour can negatively impact market liquidity and reduce the informativeness of prices regarding asset fundamentals. Furthermore, the focus of HFT on short-term technical arbitrage can decouple asset prices from a company's long term fundamental value, challenging the notion of allocative efficiency—the market's ability to direct capital to its most productive uses. When asset prices start to focus on reflect short-term trading dynamics rather more than long-term valuation, the primary role of securities markets in efficient capital allocation is challenged.

3.5 The Democratization of Wealth Management: Robo-Advisors and the Assault on High Fees

3.5.1 The Financial Imperative: Why Robo-Advisors Thrive in India

The wealth management landscape in India, like global markets, has historically been dominated by a high-cost structure. Investors seeking advice often are made to choose from the following options:

1. **Active Mutual Funds**: Run by active fund managers and have higher total expense ratios (typically 1–2%) based on the justification of superior manager skills to pick the right assets to beat the benchmark index and create alpha.
2. **Portfolio Management Services (PMS)**: Only cater to High-Net-Worth Individuals (HNIs) with minimum investments often of at least INR 50 lakhs. Most PMS funds charge a high fixed fees (e.g. 2%) as compared to mutual fund. Further, many PMS funds also charge a substantial performance incentive e.g. 20% variable over a hurdle rate (Aggarwal and Vadgama 2022).

There is growing empirical evidence, however, suggests that that higher TER charges in active funds are questionable. Majority of the active fund often fail to consistently beat their benchmarks, especially after factoring in high expense ratios (Bogle 2016; Crane and Crotty 2018). The advent of Robo-Advisors provides a technological disruption to this structural inefficiency and may benefit the end investors.

Robo-Advisor refers to automated support for all financial advisory services without any human intervention," operating on a pre-defined algorithm to calculate optimal returns and plans based on data collected from the investor. The core value proposition of these platforms is built on: (1) **Cost-Effectiveness and Scalability**: Leveraging technology to provide low-cost advice accessible to the mass market with low or zero minimum account balances. (2) **Disciplined Investing**: Utilizing online surveys for risk profiling and focusing portfolio construction heavily on low-cost Passive Mutual/Index Funds to minimize cost leakage. (3) **Automation Without Human Bias**: Continuous, emotionless portfolio management without inherent human behavioural biases through services like constant rebalancing to maintain the target asset allocation (e.g. 60% equity, 40% debt) based on the investor's profile and risk-taking capacity and intentions.

3.5.2 Quantifying the Cost of Fees: An Empirical Analysis of Wealth Erosion

To fully understand the disruptive potential of the low-cost model of Robo-Advisors, it is essential to quantify the opportunity cost of high fees over a long investment horizon. This illustration models the long-term retirement value for an individual making a moderate annual investment over a 40-year career, a timeframe typical for retirement savings.

- **Annual Investment (P)**: INR 10,000 (Invested at the beginning of each year).
- **Investment Horizon (n)**: 40 years.
- **Gross Annual Return (g)**: 10.0%.

The calculation of the closing retirement corpus uses the Future Value of Annuity Due formula, given the investment is made at the beginning of each year:

Analysis of Fee Impact: The results in Table 3.1 depict the significant long-term impact of even seemingly minor percentage differences in fees, due to the compounding over decades. The marginal cost of an additional 2.4 percentage points in annual fees (Case 3 versus Case 1) results in an opportunity cost of over INR 23.11 lakhs—representing a loss of approximately 50% of the potential final corpus. Of course, it is important to note that this can differ based on the TER of your active fund and trail commissions.

This provides justification for the disruptive value proposition of low-cost, technology-driven financial advice, affirming the need for SEBI's focus on transparency and the growth of the Robo-Advisory segment. It highlights that in the long run, the actual 'alpha' for the retail investor is often found not in stock picking, but in fee minimization. Even the legendary Warren Buffett has argued that it is difficult for active fund management to beat passive fund management due to the laws of mathematics.

Table 3.1 Comparative impact of advisory fees on long-term retirement wealth

Fee scenario	Annual fee (F)	Net annual return (%)	Closing retirement corpus (INR)	Opportunity cost (versus case 1)
Case 1: low-cost index fund (robo-advisor proxy)	0.10% (expense ratio)	9.90	47,55,000	–
Case 2: high-cost active fund	1.50% (expense ratio)	8.50	30,75,000	16,80,000
Case 3: Active fund + broker trailing fee	2.50% (1.5% ER + 1.0% trailing fee)	7.50	24,43,500	23,11,500

Source Authors own calculation

3.6 Case Study of Jane Street—Regulatory Concerns

We can't deny the fact that the algorithm-based high-frequency trading has been growing in India. This is based on the fact that Algo trading plays a significant role in increasing trading volumes, which account for 57% of equity cash and 70% of F&O trades in mid-2025. Under this heightened popularity, there is a continuous scrutiny going on by regulators. The concerns become more crucial and rigour after the happening of trading activities of Jane Street, an international Algo trading firm widely known for its high-frequency and voluminous trading strategies. In a single day, voluminous trading done by the international Algo giant raised the concerns of SEBI, which made a massive abnormal profit of Rs 735 crore in a single day. This episode triggered the debate around market fairness, ethical concerns, and overall protection of retail investors. This incident reinforces that regulators should review their policies towards Algo trading. The regulators then onwards are constantly coming up with various mechanism such as registration of Algo traders, providing unique Algo IDs, auditing of Algo traders and their due diligence, and approval of Algo strategies. These measures ensure the inclusivity and equal treatment of all participants.

3.7 Regulatory Framework: Mitigating Automation Risk in Indian Financial Markets

Considering the evolving risks with these technological advancements in financial markets, SEBI is also making constant changes and improvements in the regulatory framework to safeguard the investor.

3.7.1 Circular on Algorithmic Trading

The initial master circular was issued by SEBI on March 30, 2012, on Algo trading. As discussed earlier, the share of algorithmic trading in the Indian financial market has expanded, and more guidelines have been issued by SEBI.

SEBI, in its order dated Sept 02, 2022, made sweeping changes to reduce mis-selling and protect the interests of investors from algorithmic trading. The brokers dealing with algorithmic trading were now mandated not to indulge in the following:

- 'Directly or indirectly make any reference to the past or expected future return/performance of the algorithm; and/or'
- 'Directly or indirectly associate with any platform providing any reference to the past or expected future return/performance of the algorithm'.
- 'Stockbrokers who are directly/indirectly referring to any past or expected future return/performance of an algorithm or are associated with any platform

providing such reference, shall remove the same from their website and/or dissociate themselves from the platforms providing such references, as the case may be, within seven days from the date of this circular'.

In its recent circular issued in February 2025, the Securities and Exchange Board of India (SEBI) has taken significant steps to enhance accountability in algorithmic trading (AT). This move is particularly important given the historical complexity and anonymity that have often shielded high-frequency traders from scrutiny.

3.8 The Emergence of the Hybrid Model

The future of financial advisory is not about pitting man against machine; rather, it lies in the collaborative synergy of the Hybrid Advisory Model. This innovative approach merges the technological advantages of Robo-Advisors such as their scalability, low-cost structure, and disciplined execution with the invaluable qualities of human financial planners, including complex problem-solving abilities, emotional intelligence, and the capacity to build trust.

In this model, human advisors are liberated from the repetitive and time-consuming tasks that AI can handle efficiently. This allows them to concentrate on high-value activities that truly enhance the client experience. For instance, they can engage in complex goal setting, helping clients articulate their long-term financial aspirations and develop tailored strategies to achieve them. Additionally, human advisors can provide behaviour coaching, guiding clients through the emotional ups and downs of investing and helping them maintain a disciplined approach during market fluctuations.

Tax optimization is another area where human expertise shines. While Robo-Advisors can offer basic tax strategies, a human advisor can navigate the intricacies of tax laws and regulations, providing personalized advice that maximizes tax efficiency based on an individual's unique financial situation.

By leveraging the strengths of both technology and human insight, the Hybrid Advisory Model creates a more holistic and effective financial advisory experience. Clients benefit from the efficiency and cost-effectiveness of Robo-Advisors while also receiving the personalized support and guidance that only a human advisor can provide. This collaborative approach not only enhances client satisfaction but also fosters long-term relationships built on trust and understanding.

As the financial landscape continues to evolve, embracing this hybrid model will be essential for advisors looking to meet the diverse needs of their clients. By combining the best of both worlds, the financial advisory industry can create a future that is not only more efficient but also more empathetic and responsive to the complexities of individual financial journeys.

3.9 Conclusion

The chapter is an attempt to capture how AI and ML are changing the financial sector, particularly trading and financial advisory verticals. These two disruptive technologies have given rise to the two platforms, i.e. Algo trading used by traders trading in the stock market and Robo-Advisors in financial planning. The study examines both platforms through the lens of financial and technology adoption theories such as Efficient Market Theory, Behavioural Finance Theory, Innovation Diffusion Theory, and Principal-Agent Theory. It has been found that the hypothesis derived from EMH has been rejected by the Algo trading platform because the High-frequency trading (HFT) is generating abnormal returns by using statistical and time series techniques on the past data. The strategies have been tested on historical data by implementing back testing. Both platforms that are Robo-Advisors and Algo trading are automated and based on a set algorithm and thus free from human bias. It has been found that the adoption of technology is based on the perceived benefits, like efficiency, lower cost, and free from human bias. The technology also improved the relation between investors and intermediaries and thus principal and agents having no conflicting interest.

The study provided the conceptual framework of how AI and ML technology provided two platforms and having their own benefits and challenges; however, the most important challenge is market instability that triggers in India from the episode of Jane Street. This makes regulators active and prompted to take regulatory measures like Algo ID, API restrictions, kill switch, and Audit trails.

Finally, the study suggested that the future of financial advisory lies in the **Hybrid Advisory Model**, where human expertise and robo-technology work in partnership. This model combines the efficiency, scalability, and cost advantages of Robo-Advisors with the emotional intelligence and strategic thinking of human advisors. By automating routine tasks, advisors are able to focus on personalized goal setting, behavioural coaching, and advanced tax planning. This collaborative approach delivers a more holistic, trusted, and client-centric financial experience.

References

Aggarwal V, Vadgama C (2022) Portfolio management services in India: returns and incentives. Univers J Account Financ 10(2):476–482

Aggarwal V, Batra S, Yadav M, Kumar P (2023) Literature review on algorithmic trading in financial markets. In: 2023 International conference on sustainable Islamic business and finance (SIBF), pp 76–80. https://ieeexplore.ieee.org/abstract/document/10379947/

Akansu AN (2017) The flash crash: a review. J Cap Mark Stud 1(1):89–100

Bhatia A, Chandani A, Chhateja J (2020) Robo advisory and its potential in addressing the behavioral biases of investors—a qualitative study in Indian context. J Behav Exp Financ 25:100281

Bogle JC (2016) The index mutual fund: 40 years of growth, change, and challenge. Financ Anal J 72(1):9–13. https://doi.org/10.2469/faj.v72.n1.5

Chandani A, Sriharshitha S, Bhatia A, Atiq R, Mehta M (2021) Robo-advisory services in India: a study to analyse awareness and perception of millennials. Int J Cloud Appl Comput (IJCAC) 11(4):152–173

Crane AD, Crotty K (2018) Passive versus active fund performance: do index funds have skill? J Financ Quant Anal 53(1):33–64

Dhingra B, Batra S, Aggarwal V, Yadav M, Kumar P (2023) Stock market volatility: a systematic review. J Model Manag. https://www.emerald.com/insight/content/doi/10.1108/JM2-04-2023-0080/full/html

Dubey RK, Babu AS, Jha RR, Varma U (2022) Algorithmic trading efficiency and its impact on market-quality. Asia-Pac Finan Mark 29(3):381–409

Dubey RK, Chauhan Y, Syamala SR (2017) Evidence of algorithmic trading from Indian equity market: interpreting the transaction velocity element of financialization. Res Int Bus Financ 42:31–38

Fama EF (1970) Efficient capital markets: a review of theory and empirical work. J Financ 25(2):383–417

Fatima S, Chakraborty M (2024) Adoption of artificial intelligence in financial services: the case of robo-advisors in India. IIMB Manag Rev 36(2):113–125

Illiashenko P (2017) Behavioral finance: history and foundations. Visnyk Natl Bank Ukr 239:28–54

Kesharwani S, Prakash A, Gangwar JD (2024) Robo-advisors: automated algorithm-driven wealth management services-a literature review. Glob J Enterp Inf Syst 16(2):91–99

Kirilenko A, Kyle AS, Samadi M, Tuzun T (2017) The flash crash: high-frequency trading in an electronic market. J Financ 72(3):967–998

Malkiel BG, Fama EF (1970) Efficient capital markets: a review of theory and empirical work. J Financ 25(2):383–417

Marketsandata (2025) Markets and Data. https://www.marketsandata.com/industry-reports/india-algorithmic-trading-market

Moneycontrol (2025) Algos are now the dominating force in Indian equities; F&O share in derivatives nearly 70%. https://www.moneycontrol.com/news/business/markets/algos-are-now-the-dominating-force-in-indian-equities-f-o-share-in-derivatives-nearly-70-13086705.html

Nahar J, Nishat N, Shoaib ASM, Hossain Q (2024) Market efficiency and stability in the era of high-frequency trading: a comprehensive review. Int J Bus Econ 1(3):1–13

Nain I, Rajan S (2023) Algorithms for better decision-making: a qualitative study exploring the landscape of robo-advisors in India. Manag Financ 49(11):1750–1761

Parashara N, Sharmab R, Joshia A (2024) Trust, tech and tradition: a multifactor analysis of Indian investors' intentions to adopt Robo advisors. Glob Bus Financ Rev 29(11):128

Ramkumar GOWTHAM (2018) A study on the significance of algorithms tradings in indian stock market. Int J Res Anal Rev (IJRAR) 5

Riyazahmed DK (2024) Algorithmic trading models, aversion, and the Indian scenario. Contemp Res Manangement (CRM)

Rogers EM, Cartano DG (1962) Methods of measuring opinion leadership. Public Opin Q 26:435–441

Sharma S, Tiwari AK, Nasreen S (2022) Are FinTech, robotics, and blockchain index funds providing diversification opportunities with emerging markets? Lessons from pre and postoutbreak of COVID-19. Electron Commer Res:1–30

Vaibhav Aggarwal works as an Associate Professor at Jindal Global Business School, O.P. Jindal Global University, Haryana, India. He has interest in the areas of financial markets. He teaches subjects like financial modelling, corporate finance. He has published more than 40 scopus indexed papers.

Mahender Yadav works as an Assistant Professor at Maharshi Dayanand University, Haryana, India. He is a keen researcher and has published more than 35 papers in scopus.

Shashank Sharma is a seasoned Product Management Executive at JP Morgan Chase & Co., USA, with over 18 years of experience in data management, machine learning operations, and business process optimization. He leads a team of product managers, delivering user-centric solutions and ensuring compliance with regulatory standards.

Dr. Reepu is currently associated with Chandigarh University, India, where she contributes significantly to academia and research. She holds an MBA in Finance and a PhD in Finance, reflecting her strong academic background. With over a decade of extensive academic and research experience, she has made notable contributions to both local and international research organizations. Dr. Reepu has presented her work at esteemed conferences in India and abroad. Her core areas of expertise include Behavioral Finance, Business Management, and Financial Management. Passionate about teaching and research, she strives to inspire students and foster innovation in her field.

Comparative Analysis of Volatility in Green Cryptocurrencies Versus Conventional Cryptocurrencies

4

Indira Bhardwaj and Reepu

Abstract

By introducing decentralised financial systems as a critical factor, cryptocurrencies have altered and enhanced the financial industry dramatically. Despite the growth in popularity, the extreme price fluctuations of cryptocurrencies have been identified as a vital issue. Furthermore, since traditional cryptocurrencies like Bitcoin were based on high-energy-consuming tasks, green cryptocurrencies with energy-saving consensus mechanisms are evolving. The information about conventional cryptocurrencies is enough, but not much is there about green cryptocurrencies' behaviour of high volatility, nor are quantitative measures there. The purpose of this research is to contrast the volatility profiles of green and conventional cryptocurrencies with the help of different econometric techniques. For instance, the Generalized Autoregressive Conditional Heteroskedasticity (GARCH 1, 1) model is deployed to check for the existence of volatility clustering and the Granger Causality test is used to ascertain potential correlation between the two categories of cryptocurrency. The research not only identifies the volatility of cryptocurrencies but also offers valuable information for the environmentally conscious investor who is also an investor in digital assets. These sustainable cryptocurrencies are the subject of the present study that through the analysis of their volatility dynamics it can be possible to find out how they behave in the crypto market.

I. Bhardwaj (✉)
Maharaja Agrasen Technical Education Society, New Delhi, India
e-mail: indira12jan1973@gmail.com

Reepu
Chandigarh University, Mohali, Punjab, India

B. Kaur et al. (eds.), *Integrating Industry 5.0 and 4.0 for Technology and Humanity*,
Contributions to Environmental Sciences & Innovative Business Technology,
https://doi.org/10.1007/978-981-92-1780-9_4

Keywords

Green cryptocurrencies • Conventional cryptocurrencies • Granger causality • GARCH (1, 1) • Volatility

4.1 Introduction

Cryptocurrencies have come to be a notable part of today's financial markets, presenting a decentralised alternative to the traditional banking systems (Nakamoto 2008). As far as their quick usage, they have caused the emergence of new possibilities; however one of the most debated aspects is the extreme price volatility of cryptocurrencies (Aggarwal et al. 2023; Yadav et al. 2023; Bouri et al. 2017). Though most people consider cryptocurrencies as volatile, their volatility remains a debatable issue. Most likely, it benefits risk-addicted traders, but it also makes them very unsteady and of lesser use as financial assets in the long run (Corbet et al. 2018). Apart from the financial risks, the environmental impact of cryptocurrencies, in particular those that rely on the energy-intensive Proof-of-Work (PoW) method, has received immense attention. These systems need enormous computational power and electricity consumption that are often derived from non-renewable energy sources. Consequently, they contribute a lot to carbon emissions and ecological degradation.

On the other hand, green cryptocurrencies have made their area a gainer as they can use consensus mechanisms such as Proof-of-Stake (PoS) to not dirty the earth with their carbon footprints (Sedlmeir et al. 2020). Green cryptocurrency, a term coined for environmentally friendly cryptocurrencies, employs consensus mechanisms that are energy-efficient and produce the least environmental damage. For instance, they would use a protocol called Proof-of-Stake (PoS), a mechanism termed Delegated Proof-of-Stake (DPoS) or the latter but one. Thus, the networks of the green cryptocurrencies will not only be energy-efficient but also neutral or negative in terms of their environmental imprint (Sedlmeir et al. 2020). Cardano (ADA), Algorand (ALGO) and Hedera (HBAR) can be mentioned as the fair new cryptocurrencies taking the path of aligning their development with global sustainability goals of all countries. The research into the different types of well-known cryptocurrencies volatility patterns (Bitcoin and Ethereum) is quite abundant, while the knowledge about green cryptocurrencies is very limited. These digital currencies meet the sustainability objectives. Hence, it is of immense importance to verify whether their price ascertains are different from the regular cryptocurrencies.

The main point of the query is how the eco-character of green cryptocurrencies affects the investors' emotions and thus decreases the volatility of the investors. The differentiations between these are important as investors are increasingly incorporating environmental, social and governance (ESG) factors into their decision-making (Truby and Brown 2021). This research derived the volatility of green and conventional cryptocurrencies from a number of econometric techniques such as the Generalized Autoregressive Conditional Heteroskedasticity (GARCH) model were used to identify volatility clustering, further Granger Causality was employed to check interdependencies, the Augmented Dickey-Fuller (ADF) test was executed to affirm stationarity and to describe market behaviour (Dickey and Fuller 1979). By introducing the above-mentioned models, this article endeavours to give a better understanding of the 'risk and volatility' themes of green cryptocurrencies when compared to regular cryptos. The results would create a new field of studies about the ecological finance and assist the investors and decision-makers to form a sustainable digital asset strategy.

4.2 Literature Review

In recent years, the volatility of cryptocurrencies has gained momentum and is under discussion of academicians, practitioners and policymakers. This has been backed by the surging prices, volatility and diversification benefits. From the advent of cryptocurrency in 2015 to the present, we have seen several phases. It was believed that cryptocurrencies were considered a speculative asset class. The investors considered cryptocurrencies as a diversification asset alongside other non-conventional asset classes. There are strands of literature that empirically analysed that cryptocurrencies are considered as a safe hedge against volatility and provide diversification with other asset classes (Bhuiyan et al. 2025; Nguyen 2025; He et al. 2024; Liu and Yuan 2024; Baur et al. 2022; Urquhart and Zhang 2019). However, some studies found that the cryptocurrencies are now aligned with fundamentals and shown co-movement with the stock market and other financial asset (Adelopo and Luo, 2025; Hu et al. 2023; Kumah et al. 2022).

Over time, cryptocurrencies have been criticised because of environmental concerns, primarily the huge energy consumption of Proof-of-Work (PoW), which triggers an environmentally unfriendly asset class. In this context, several digital asset classes emerged, such as green funds/ETFs. These funds are based on a green/renewable theme and thus target companies that are linked to green investments. The studies found that green ETFs, as a non-conventional are providing diversification benefits (Antar et al. 2025; Rizvi et al. 2022). Under green thematic investment, another cryptocurrency emerged that runs on proof of stake (PoS), and some tokens are linked to green investments. Ul Haq and Bouri (2025) had analysed the volatility of green crypto; some studies (Zuo et al. 2025; Chen and Nguyen 2024; Lobao 2022) analysed the herd behaviour of green crypto/tokens. Furthermore, there are strands of literature found the spillover to other asset class and hence proved to be a good diversifier. Few studies (Danila 2025; Kaur et al. 2025; Naeem et al. 2023) contradict the hypothesis that green crypto are safe hedge during volatility and provide diversification.

It has been found that the majority of the literature is skewed towards volatility and diversification of green crypto/tokens. However, the comparative study of green and conventional crypto, only few studies has been done by Bhattacharya et al. (2025), Duan et al. (2023), Haq and Bouri (2022). The studies found week correlation among green and conventional crypto. Dias et al. (2025) have analysed the comparative efficiency of conventional crypto and green crypto. By analysing the strands of literature, we have found that the causal movement and comparative volatility still debatable and thus required an attempt to validate the results and value to the ongoing discussion. This study extends the discussion by capturing the behaviour of conventional and crypto, understanding their volatility by the application of econometric-based volatility models, and additionally, the study captures the directional causality. The study provides relevant insights to investors and policymakers.

4.3 Data Profile and Methodology

The study has applied combination of statistical and econometric models to capture the behaviour, static volatility, causality and time varying volatility of conventional and green cryptocurrencies. This has been done to investigate the similarity and dissimilarity among two sets of cryptocurrency. The methods and models are applied in the study have been started from preliminary analysis to capture the comparative behaviour, followed by Augmented Dickey-Fuller (ADF) test, to comment on its stationarity of returns at its level. Furthermore, to capture the causal movement, Granger Causality test has been applied to conclude that the causality movement is whether no directional, uni-directional or no directional. Finally, to capture time varying volatility, the study applied symmetric GARCH (1.1) model, to understand the short term and long term volatility. All the insights that have been derived at various stage will provide insights to investors and contributing to literature.

The study has considered BTC-USD (Bitcoin), ETH-USD (Ethereum) and USDT-USD (Tether), as a proxy of conventional asset class. On the other hand, ADA-USD (Cardano) and ALGO-USD (Algorand), XCH-USD (Chia) are considered as the proxy of green crypt currencies. These crypto are selected on the basis of their market capitalisation. The daily prices are fetched from www. Investing. com, from January 2022 to March 2025. To prepare the data, daily returns were calculated using the formula:

$$\text{Return} = \ln(\text{Pt}/\text{Pt} - 1), \tag{4.1}$$

where *Pt* is the price on the current day, and *Pt-1* is the price on the previous day. This log-return approach helps in making the data more normally distributed and suitable for statistical analysis. The methodology has been discussed sequentially.

4.3.1 Preliminary Analysis

The analysis captures the behaviour of conventional and environmental friendly currencies by computing mean, median, minimum, maximum, standard deviation, skewness and kurtosis. The analysis prima facie is able to capture the returns, volatility and probability of getting positive of negative returns and finally, peakedness of the returns will add value to the analysis. This provide specific strategy to investors and also help to provide detail understanding that a particular asset would be suitable to which type of investor.

4.3.2 Augmented Dickey-Fuller Test

The test is used to analyse the stationarity of the asset class. It has been seen that the returns of the assets are varying according to the time and thus having leverage effect. The series is said to be stationarity, if its mean, variance and correlation are constant. The data is desirable to be stationary for any econometric model, otherwise the coefficients derived from the model become spurious and not stable. Henceforth, stationarity is the critical assumption for the application of mean reverting models like VAR, ARIMA and GARCH. The Augmented Dickey-Fuller is designed to handle serial correlation in the error terms by including lagged differencing of the dependent variable. The ADF regression equation can be expressed as:

Here,

$$\Delta y_t = \alpha + \beta_t + \partial Y_{t-1} + \sum_{i=1}^{p} \delta \Delta y Y_{t-1} + \varepsilon t. \tag{4.2}$$

Δy_t represents the first difference of the series, tis the time trend, Y_{t-1} is the lagged level of the series, and p is the number of lagged difference terms included to eliminate autocorrelation.

4.3.3 Granger Causality Test

The genesis of diversification is based on low correlation among two asset class. In financial arena, the causality movement provides better understanding to capture the lead and lag analysis. It has been believed that giver is better that receiver. Thus, the directional causality serves value understand the genesis of volatility and how this volatility spread in the entire market. One of econometric model to capture the directional volatility is Granger Causality (Granger 1969). The model captures. The causal relation can be defined as *X* causes *Y* if the past behaviour of the time series of *X* series able to capture the behaviour of *Y* series. The peculiarity of the Granger Causality is that the model is able to identify the causal relation between two time series without having any a priori hypothesis. Causality captures in three ways-no directional, uni-directional and bi-directional. Granger Causality equation can be written as follows:

$$X_{(t)} = \sum_{j=1}^{p} A_{11,j};\ X_{(t-j)} + \sum_{j=1}^{p} A_{12,j};\ Y_{(t-j)} + \varepsilon_{1(t)}, \tag{4.3}$$

$$Y_{(t)} = \sum_{j=1}^{p} A_{21,j};\ X_{(t-j)} + \sum_{j=1}^{p} A_{22,j};\ Y_{(t-j)} + \varepsilon_{2(t)}, \tag{4.4}$$

where p denotes the lagged of observations included in the model. The study captures the directional causality among conventional and green cryptos, to understand their behaviour, and insights derived from the analysis are useful for investors and policymakers.

4.3.4 GARCH (1, 1)

GARCH model is an extension of ARCH given by Bollerslev (1986). σ_t^2 is calculated through GARCH model which is called conditional volatility. The given condition of series Y_t is represented as follows:

$$Y_t|\varphi_{t-1} \sim N\left(0, \sigma_t^2\right).$$

σ_t^2 shows conditional volatility while φ_{t-1} depicts the tangible news at time t.

The suitable GARCH (1, 1) model is shown as follows:

$$\sigma_t^2 = \alpha + \beta_1 \in_{t-1}^2 + \beta_2 \sigma_{t-1}^2 + \mu_t. \tag{4.5}$$

Further, GARCH (p, q) model can be written as:

$$\boldsymbol{\sigma_t^2 = \alpha + \sum_{i=1}^{p} \beta_1 \in_{t-1}^2 + \sum_{j=1}^{q} \beta_2 \sigma_{t-1}^2 + \mu_t},$$

where

$$\alpha > 0;\ \beta_1 \geq 0;\ \beta_2 \geq 0.$$

GARCH (1, 1) model indicates that the conditional volatility of any series depends on lagged squared residuals $\in_{t-1}^2$ while value σ_{t-1}^2 i.e. GARCH effect.

The rationality in applying the model is to capture the volatility persistence. It helps to capture the conditional volatility of a given series. Generally, it is popular in terms of standard GARCH which is one of the models to measure symmetrical volatility of a series. It is the sum of the ARCH term and GARCH term.

4.4 Empirical Analysis

4.4.1 Discussion on Descriptive Statistics

Table 4.1 captured the preliminary analysis of the overall behaviour of conventional and green cryptocurrencies. The study considered three conventional such as Bitcoin (BTC-USD), Ethereum (ETH-USD), Tether (USDT-USD) and three green crypto such as Cardano (ALGO-USD), Algorand (ADA-USD) and Chia (XCH-USD). The analysis provides the return and risk profile of select asset class; however, the risk captured here is static rather dynamic.

This analysis highlights that Bitcoin and Ethereum, among conventional asset class, are providing positive returns. Thus, these two are providing attractive returns, whereas Tether is behaving like a stable asset throughout the period and displays a mean return as zero and with lowest volatility. This reveals that among three crypto, Tether is suitable for conservative investors. On the other hand, the green cryptocurrencies such as Cardano, Algorand and Chia, Algorand and Chia both are providing positive, whereas Cardano is fetching negative mean return.

Taking into consideration of volatility of the cryptocurrencies, among conventional cryptocurrencies, BTC is negatively skewed (− 0.168) and whereas Ethereum is positively skewed (0.221), with high kurtosis in their returns. In contrast Tether exhibits positive skeweness with comparatively low kutosis. It means that Ethereum is an attractive asset, with positive mean return with high probability of fetching positive returns. The findings suggest that both Etherum and Bitcoin may be suitable for aggressive investors, Tether considered a stable and could be best suited for hedgers.

Considering green cryptocurrencies, Cardeno is exhibiting negative mean returns, whereas both Algorand and Chia are fetching positive returns. Comparative to conventional cryptcurrencies, green cryptocurrencies are exhibiting high Kurtosis ranging from 7.402 to 21.219. We can conclude that the green cryptocurrencies, although more volatile, could offer rewarding opportunities to the investors and thus suitable for aggressive investors.

Table 4.1 Descriptive statistics

Statistics	BTC	ETH	USDT	ALGO	ADA	XCH
Mean	0.000	− 0.001	0.000	− 0.002	0.000	0.002
Median	0.000	0.000	0.000	0.001	− 0.001	− 0.003
Std dev	0.028	0.086	0.001	0.049	0.046	0.042
Variance	0.001	0.007	0.000	0.002	0.002	0.002
Min	− 0.174	− 0.192	− 0.004	− 0.273	− 0.278	− 0.259
Max	0.136	0.176	0.005	0.308	0.538	0.447
Skewness	− 0.168	0.221	0.503	− 0.082	1.227	1.751
Kurtosis	7.246	7.164	3.578	7.407	21.186	21.219

Source Owner's own work

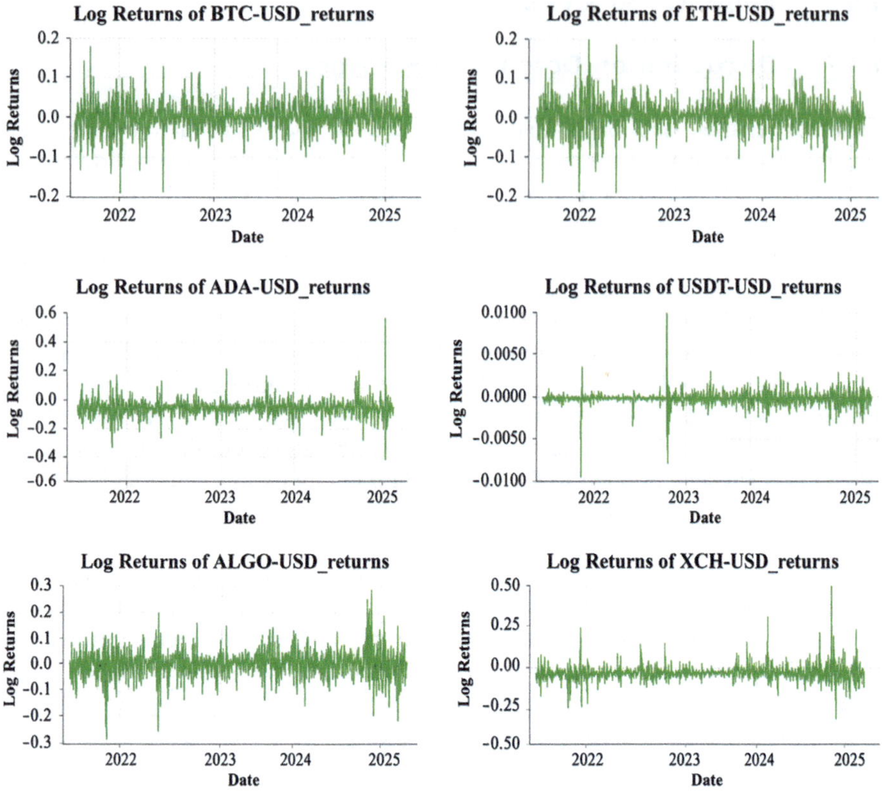

Fig. 4.1 Log returns of conventional and non-conventional cryptocurrencies. *Source* Owner's own work

Figure 4.1 captures the time-varying volatility of conventional and green cryptocurrencies and highlights the presence of, leverage effects and volatility clustering. The key observation is that the time-varying patterns of these two categories of cryptocurrencies differ considerably, exhibiting distinct dynamics. Thus, from the preliminary observation, we can conclude that green cryptocurrencies could provide diversification benefits to conventional crypto currencies investors. However, this could be further investigated from other econometric models

4.4.2 Discussion on ADF Test

Table 4.2 presents the results of the Augmented Dickey-Fuller (ADF) test, which is used to assess the stationarity of each cryptocurrency's price series. Understanding stationarity is crucial for making accurate time-series forecasts.

The findings reveal that the raw price series for Bitcoin, Ethereum and Cardeno are non-stationary. This is indicated by test statistics that exceed critical values

Table 4.2 Results of ADF test

Cryptocurrency	Dickey-fuller statistic	*P*-value	Stationarity
BTC-USD	– 2.28	0.460968	Not stationary
ETH-USD	– 2.49	0.368847	Not stationary
USDT-USD	– 6.30	0.01	Stationary
ALGO-USD	– 5.76	0.01	Stationary
ADA-USD	– 2.73	0.267726	Not stationary
XCH-USD	– 3.53	0.039153	Stationary

Source Owner's own work

and p-values greater than 0.05. This non-stationarity is consistent with their speculative nature and their sensitivity to external market influences, which can lead to unpredictable price movements. In contrast, Tether, Algorand, and Chia show stationarity at their level, meaning they tend to revert to a mean value and are less susceptible to random fluctuations. This characteristic makes them more stable compared to the other cryptocurrencies.

However, when we apply first-order differencing to all cryptocurrencies, they all become stationary. This transformation is significant because it prepares the data for econometric models like GARCH and Granger Causality, which require stationarity as a fundamental assumption. With the data now well-conditioned, we can effectively model shocks, analyse volatility clustering and forecast future behaviour in the cryptocurrency market.

4.4.3 Discussion on Granger Causality Test

Table 4.3 presents the results of the Granger Causality tests, which examine whether the past returns of one cryptocurrency can predict the returns of another. The test has been applied among conventional and green cryptocurrencies. The output of the tests among various combinations is providing very interesting results to investors, especially to the investors investing in conventional cryptocurrencies. The results shows that both are operating in a peculiar way, there are no signs of spillover or contagion effect, and the results are resonating the earlier results. It has been concluded that conventional cryptocurrencies are not a cause of green cryptocurrencies, and on the other hand, green cryptocurrencies are also not a cause of the volatility perceived in conventional. Because of their peculiar behaviour, finally it can be hypothesised that there is no cause and effect relationship exist among two. The results are very interesting for investors investing in cryptocurrencies and other asset class.

Table 4.3 Results of granger causality test

Cause	Effect	Statistic value	*P*-value	Significance
Bitcoin	Cardeno	0.4513	0.6369	Not significant
Bitcoin	Algorand	1.9191	0.1472	Not significant
Bitcoin	Chia	1.2005	0.2755	Not significant
Ethereum	Cardeno	0.4802	0.6188	Not significant
Ethereum	Algorand	0.8111	0.4446	Not significant
Ethereum	Chia	1.8421	0.1589	Not significant
Tether	Cardeno	1.349	0.259	Not significant
Tether	Algorand	1.4911	0.2905	Not significant
Tether	Chia	0.6749	0.5004	Not significant

Source Owner's own work

4.4.4 Discussion on GARCH Model Analysis

Table 4.4 delves into the volatility dynamics of the cryptocurrencies using a GARCH (1, 1) framework, providing valuable insights into their behaviour in the market. Starting from the mean return, it has been found that green cryptos are again outperforming in comparison to conventional cryptocurrencies. Among all the cryptocurrencies considered, Chia consistently exhibits superior performance, the results are in consensus across the analysis, especially with the results of preliminary analysis. The ARCH (α) term represents short term volatility, if we can say that Tether is again showing stability in short term among conventional cryptocurrencies, followed by Bitcoin and Ethereum. The results are in consonance with preliminary analysis. In contrast to conventional cryptocurrencies, Cardeno, Algorand and Chia are exhibiting high short term volatility. The results are particularly relevant for traders, as they may provide opportunities to benefit from intraday trading.

In the context of long term volatility captured through GARCH (β), these two distinct fraternity are showing peculiar long term trends. The distinct behaviour of both the types of assets is providing diversification opportunities. Considering overall volatility, which reflects the combined effects of short and long-term

Table 4.4 Results of GARCH model (1, 1)

Cryptocurrency	Mean return	ARCH (α)	GARCH (β)	Log likelihood	Overall volatility
BTC-USD	0	0.05	0.81	7574.31	0.86
ETH-USD	0	0.092	0.82	2318.545	0.912
USDT-USD	0.001	0.173	0.624	2582.673	0.797
ALGO-USD	0.002	0.153	0.798	2004.696	0.951
ADA-USD	0.001	0.173	0.791	2050.77	0.964
XCH-USD	0.008	0.198	0.786	2223.849	0.984

volatility, Tether consistently exhibits greater stability, whereas Ethereum emerges as the most volatile cryptocurrency among the conventional cryptocurrencies. The green cryptocurrencies are showing high total volatilities compared to conventional. Among green crypto, Chia is the most volatile and thus considered to be suitable for aggressive investors.

Overall, we can summarise that the environmental-based crypto are showing distinct behaviour from conventional crypto. They are suitable for short-term trading, lucrative for investments, hedgers can adopt diversification strategies to hedge their existing position, while these cryptocurencies may also be suitable for aggressive investors.

Source Owner's own work.

4.5 Conclusion

The study is an attempt to capture the distinct behaviour of two sets of cryptocurrencies that is conventional and green cryptocurrencies. We have considered three crypto in each of these categories. To capture their behaviour, causal relations and finally volatility, we applied preliminary analysis, ADF test, Granger Causality and finally GARCH (1, 1). We found that two set of digital currencies are showing distinct behaviour in terms of returns and static volatilities. Tether has been considered as a stable one, and Ethereum is outperforming among conventional crypto. In contrast of the conventional, green crypto are outshining and shown promising returns with high probabilities of getting outperformed returns. Chia among green crypto is considered as most outperformed digital currency.

We have also found that there is no causality among these two sets of crypto currencies that strengthen our hypothesis that these two are peculiar and shown insignificant signs of contagion behaviour. This insight is very useful for investors and hedgers. Thus, we the results of Granger Causality is showing that there is cause and effect relation among them and thus shown diversification opportunities.

Finally, the results of GARCH (1, 1) strengthen our analysis in many ways and provide interesting insights to investors. The analysis concludes that the green crypto are good for intraday trading, and the traders can generate alpha returns from the market. The analysis also provides the signs of diversification as the volatility patterns are not similar among the two sets of assets, thus strengthening the results of Granger Causality. As far as overall volatility is concerned, aggressive investors can take the benefit of high volatility and can thus get profits from short and long term position. Among all three green cryptocurrencies, Chia standout due to its high return, higher probability of generating positive returns, potential to serve as a safe-haven hedge, and attractiveness to both traders and investors. The study is relevant to institutional investors, regulators and policymakers.

4.6 Future Direction of the Study

Overall, the study is action driven and provides niche direction to trader and investors. However, there are certain limitations such as the study is limited to three green crypto, with such a scant analysis, we can can't generalise our hypothesis. This need to be analysed in a comprehensive way. Furthermore, there are certain methodological limitation, this study is limited to econometric model, and however, there are other sophisticated econometric models that can strengthen the analysis. To get in depth analysis and provide action-based strategies to investor, machine learning models could be implemented. In future, the study can provide well-proven strategy to investors after the application of back-testing. All this future directions will strengthen the work and provide readymade action packed strategy to investors and hedgers.

References

Adelopo I, Luo X (2025) How do cryptocurrency features determine their dynamic volatility and co-movements with stocks? Cogent Bus Manag 12(1):2461732

Aggarwal V, Sharma S, Kumar P, Mohnot J, Yadav M (2023) Lessons from the Russia-Ukraine war: portfolio optimisation in cryptos and stocks for investors in Ghana. Int J Dipl Econ 9(2):160–175

Antar M, Hassan R, Amami R (2025) Analyzing diversification opportunities among top-performing ETFs: a multi-wavelet and MGARCH-DCC approach. J Risk Financ 26(4):653–676

Baur DG, Hoang LT, Hossain MZ (2022) Is bitcoin a hedge? How extreme volatility can destroy the hedge property. Financ Res Lett 47:102655

Bhattacharya R, Mukherjee J, Roy S, Rana MT, Parveen R (2025) Eco-crypto dynamics: cointegration of green and non-green cryptocurrencies for sustainable investing. Adv Consum Res 2(3)

Bhuiyan RA, Mukherjee TC, Tarique KM, Zhang C (2025) Hedge asset for stock markets: cryptocurrency, cryptocurrency volatility index (CVI) or commodity. Quant Financ Econ 9(1):131–166

Bollerslev T (1986) Generalized autoregressive conditional heteroskedasticity. J Econ 31(3):307–327. https://doi.org/10.1016/0304-4076(86)90063-1

Bouri E, Jain A, Roubaud D, Kristoufek L (2017) Cryptocurrencies as speculative assets: risk-return characteristics and volatility dynamics. Financ Res Lett 23:1–10. https://doi.org/10.1016/j.frl.2017.03.009

Chen AS, Nguyen HT (2024) A new perspective on how investor sentiment affects herding behavior in the cryptocurrency market. Financ Res Lett 67:105737

Corbet S, Meegan A, Larkin C, Lucey B, Yarovaya L (2018) Exploring the dynamic relationships between cryptocurrencies and other financial assets. Econ Lett 173:1–5. https://doi.org/10.1016/j.econlet.2018.09.017

Danila N (2025) Connectedness of dynamic volatility over green Cryptos, the ESG indices, and gold: TVP-VAR approach. SAGE Open 15(3):21582440251343375

Dias R, Galvão R, Teixeira N, Alexandre P, Gonçalves S (2025) Efficiency of traditional and green cryptocurrencies: a comparative analysis

Dickey DA, Fuller WA (1979) Distribution of the estimators for autoregressive time series with a unit root. J Am Stat Assoc 74(366):427–431. https://doi.org/10.2307/2286348

Duan K, Zhao Y, Wang Z, Chang Y (2023) Asymmetric spillover from bitcoin to green and traditional assets: a comparison with gold. Int Rev Econ & Financ 88:1397–1417

Granger CWJ (1969) Investigating causal relations by econometric models and cross-spectral methods. Econometrica 37(3):424–438. https://doi.org/10.2307/1912791

Haq IU, Bouri E (2022) Sustainable versus conventional cryptocurrencies in the face of cryptocurrency uncertainty indices: an analysis across time and scales. J Risk Financ Manag 15(10):442

He C, Li Y, Wang T, Shah SA (2024) Is cryptocurrency a hedging tool during economic policy uncertainty? An empirical investigation. HumIties Soc Sci Commun 11(1):1–10

Hu Y, Shen D, Urquhart A (2023) Attention allocation and cryptocurrency return co-movement: evidence from the stock market. Int Rev Econ & Financ 88:1173–1185

Kaur J, Singh K, Amritha LJ (2025) Volatility spillover effect of green finance with renewable energy and cryptocurrency market. Financ Innov Glob Sustain:227–248

Kumah SP, Odei-Mensah J, Baaba Amanamah R (2022) Co-movement of cryptocurrencies and African stock returns: a multiresolution analysis. Cogent Bus Manag 9(1):2124595

Liu P, Yuan Y (2024) Is bitcoin a hedge or safe-haven asset during the period of turmoil? Evidence from the currency, bond and stock markets. Int Rev Financ Anal 96:103663

Lobao J (2022) Herding behavior in the market for green cryptocurrencies: evidence from CSSD and CSAD approaches. Sustainability 14(19):12542

Naeem MA, Nguyen TTH, Karim S, Lucey BM (2023) Extreme downside risk transmission between green cryptocurrencies and energy markets: the diversification benefits. Financ Res Lett 58:104263

Nakamoto S (2008) Bitcoin: a peer-to-peer electronic cash system. Retrieved from https://bitcoin.org/bitcoin.pdf

Nguyen BKQ, Pham DTN (2025) Investing during a Fintech revolution: the hedge and safe haven properties of bitcoin and Ethereum. Res Int Bus Financ 73:102599

Rizvi SKA, Naqvi B, Mirza N (2022) Is green investment different from grey? Return and volatility spillovers between green and grey energy ETFs. Ann Oper Res 313(1):495–524

Sedlmeir J, Buhl HU, Fridgen G, Keller R (2020) The energy consumption of blockchain technology: beyond myth. Bus Inf Syst Eng 62(6):599–608. https://doi.org/10.1007/s12599-020-00656-x

Truby J, Brown T (2021) Sustainable finance and the rise of green cryptocurrencies. J Sustain Financ Invest 11(4):456–478. https://doi.org/10.1080/20430795.2021.1915561

Ul Haq I, Bouri E (2025) Multiscale dynamic relationship between cryptocurrency implied volatility and socially responsible and green assets. J Sustain Financ Invest:1–25

Urquhart A, Zhang H (2019) Is bitcoin a hedge or safe haven for currencies? An intraday analysis. Int Rev Financ Anal 63:49–57

Yadav MP, Sharma S, Jha B (2023) Does the cryptocurrency index provide diversification opportunities with MSCI world index and MSCI emerging markets index?: cryptocurrency and portfolio diversification. In: Advanced machine learning algorithms for complex financial applications. IGI Global, pp 94–114

Zuo Y, Wang P, Duan Z, Fan H, Wu M (2025) Research on the emergence of herd behavior of tradable green certificate transaction subjects based on system dynamics. J Comb Math Comb Comput 125:7411–7433

Non-Fungible Tokens (NFTs): A Bibliometric and Systematic Review, Current Streams, Developments, and Directions for Future Research

5

Abha Gupta, Kamini Rai, and Manoj Kumar Mishra

Abstract

The present research work presents bibliometric and systematic literature review (SLR) of on the non-fungible tokens (NFTs) using the literature published in the 2018 and 2024. The aim of conducting the SLR is to identify developing themes and mark the future research directions. With the help of database accessible using Scopus, 50 papers which hold the maximum citations were analysed following the multi-level filtration process, supported by a scientometric approach to detect patterns and theoretical linkages. The review recognizes key variables, analytical tools, and methodological patterns which are commonly used in previous NFT studies, highlighting their connection with cryptocurrencies and digital assets. The findings reveal evolving research objectives and significant gaps, offering directions for future exploration. Major limitations include the use of a single database and a limited sample size, which future studies can overcome through broader longitudinal analyses. This paper contributes by enhancing understanding of NFTs' role in the digital economy and by informing policymakers about the regulatory and economic implications of NFTs. Uniquely, it employs the TCCM framework to integrate theoretical and methodological insights into NFT research.

Keywords

NFT • Cryptocurrency • SLR • TCCM • Digital assets

A. Gupta (✉) · K. Rai
Rukmini Devi Institute of Advanced Studies (RDIAS), GGSIPU, Delhi, India

M. K. Mishra
Marwadi University, Rajkot, Gujarat, India

B. Kaur et al. (eds.), *Integrating Industry 5.0 and 4.0 for Technology and Humanity*, Contributions to Environmental Sciences & Innovative Business Technology,
https://doi.org/10.1007/978-981-92-1780-9_5

5.1 Introduction

In this contemporary era, fascination about the blockchain technology among general community, policymakers, and financiers is growing rapidly (Bouraga 2021; Martha et al. 2023). Blockchain technology is supposed to be crucial technology accountable for escorting in the following stage of the Net, and it will unquestionably have a noteworthy effect on future technical progression (Arcenegui et al. 2021; Nobanee et al. 2023). Decentralized finance and tokenized funding for sustainable investments present high-risk, fringe opportunities for investors (Arora et al. 2025). It has been combined in numerous domains comprising management and economics and is usually used in the finance industry. The paramount crypto currency, Bitcoin, was introduced in 2008 under the pseudonym Satoshi Nakamoto by a unified person or group (Mochram et al. 2022; Yang 2023; Sabharwal and Mahalwala 2025). It is deposited on blockchain, an openly available distributed record. Plentiful investigations have discovered numerous rewards, features, and networks of non-fungible tokens (NFTs). NFT provides a spiteful edge tactic for exchanging digital assets and has practiced a shrill upsurge in use across a wide assortment of businesses. The blockchain's NFT, in specific, is gaining appreciation due to its flexibility in artistic sector, education, dispersed open market, and betting industries. These tokens characterize digital possessions that are considered valued and distinctive along with enabled by keen agreements and blockchain technology. Since NFT can confirm the reliable possession of digital resources within cybernetic setting and restricting transfers to NFT holders, creating it is one of the greatest noteworthy forthcoming application domains for blockchain technology and keen agreement. An extensively held fallacy is that non-fungible tokens (NFTs) were generated for gamers to trade digital goods within the gaming environment (Mochram et al. 2022; Yang 2023). The existing research covers a range of topics related to NFTs. The features of NFTs markets were examined in the first stream (Lounge 2020; Nadini et al. 2021; Jayasuriya 2023; Rahman 2023). The second part highlights the multiple benefits of NFTs such as enhanced data integrity and transparency (Cornelius 2020; Cornelius 2021; Wang et al. 2021; Doan et al. 2021; Arora et al. 2022), lower energy consumption (Valeonti et al. 2021; Popescu 2021; Truby et al. 2022; Ante 2022), enhanced security on devices (Arcenegui et al. 2020; Arcenegui et al. 2021; Rehman et al. 2021; Ferone 2022), easier payment transfer (Musan et al. 2020; Raman 2021; Popescu 2021; Ante 2023), more efficient mining and decreased harm to environment (Valeonti et al. 2021; Truby et al. 2022; Kumar 2022; Garnett 2022), wildlife protection and conservation (Mofokeng et al. 2018; Mont et al. 2023; Nogueira et al. 2023; Chalkias et al. 2024; Behl et al. 2024), improved food safety (Dos Santos et al. 2021; Turki et al. 2023; Liu et al. 2023), secured industrial supply chain software (Omar et al. 2020; Chiacchio et al. 2022), and lower incidence of crime activities (Mackenzie 2022; Das et al. 2022; Sadiqin 2023; Hufnagel 2023). The third stream explores how the costs of NFTs compare the value of various assets, such as large and financial holdings, real estate, and cryptocurrencies (Raman 2021; Kong et al. 2021; Ante 2022; Bamakan et al. 2022; Fairfield 2022).

This study makes multiple contributions to existing knowledge. The document provides a compilation of research articles about NFTs, divides the body of literature into three distinct sections, and identifies the leading nations and organizations. Furthermore, it outlines opportunities for further investigation. To the best of the authors' knowledge, there hasn't been a previous scholarly review on NFTs; thus, this work provides more valuable source of information on this subject. This is how the rest of paper is structures; in Sect. 2, the data and techniques are presented, the findings from bibliometric, systematic, and thematic analysis are interpreted in Sect. 3, the discussion on future research path is covered in Sect. 4, study limitations are discussed in Sect. 5, and summery of the findings is provided in Sect. 6.

5.2 Research Questions and Gap

The research question is designed to clarify and outline the focus of the literature review. This approach streamlines the process of data extraction and analysis, making it more efficient. In this paper, the research question centres on various issues considering the NFT studies. Following section presents the specific research questions addressed in this study.

1. What are the primary variables investigated across studies in the field of NFT? How have different studies defined and measured these variables?
2. What research methodologies (e.g. qualitative, quantitative, mixed-methods) are prevalent in the studies conducted in the past?
3. Which theoretical contexts have been used in the previously conducted research work in the domain of NFT?
4. On what research aims, previous studies were conducted and executed? What objectives, aims, and purposes are presently understated or require further exploration?

To answer the above stated research questions, the present research work performed an all-inclusive exploration of published research articles with the help of a definite search string. The search string which was utilized for exploration was 'NFT', and the related search was completed across a curated assortment of academic catalogues and journal storehouses. The approach followed for exploration ensured that the review apprehended appropriate studies on the topic of NFT (Non-Fungible Token) security, permitting a detailed investigation of prevailing research and the synthesis of findings related to this area.

5.3 Research Methodology

A systematic literature review (SLR) is defined as a planned process for studying existing research on a specific topic. The primary purpose of performing the SLR is to determine, measure, and synthesize appropriate studies to answer precise, predefined research questions. By meticulously recognizing, assessing, and interpreting research findings, this technique delivers a detailed summary of the topic in question. In the present research work, authors performed systematic literature review (SLR) to accomplish stated objectives. The foremost purpose is to recognize the key goals of previously conducted NFT studies, comprehend the crucial variables that have been assessed and analysed by previous authors, and inspect the methodological practises, tools, and techniques hired in previous investigations. Moreover, the SLR will aid the authors in pinpointing the domains/areas where further research is utmost needed. By thoroughly analysing the current available literature, authors aim to present a comprehensive detailed overview of the current position of NFT research and reveal opportunities for forthcoming exploration. The present paper adopted SLR in NFT and reviewed 50 papers for this purpose based in the flow chart represented as Fig. 5.1.

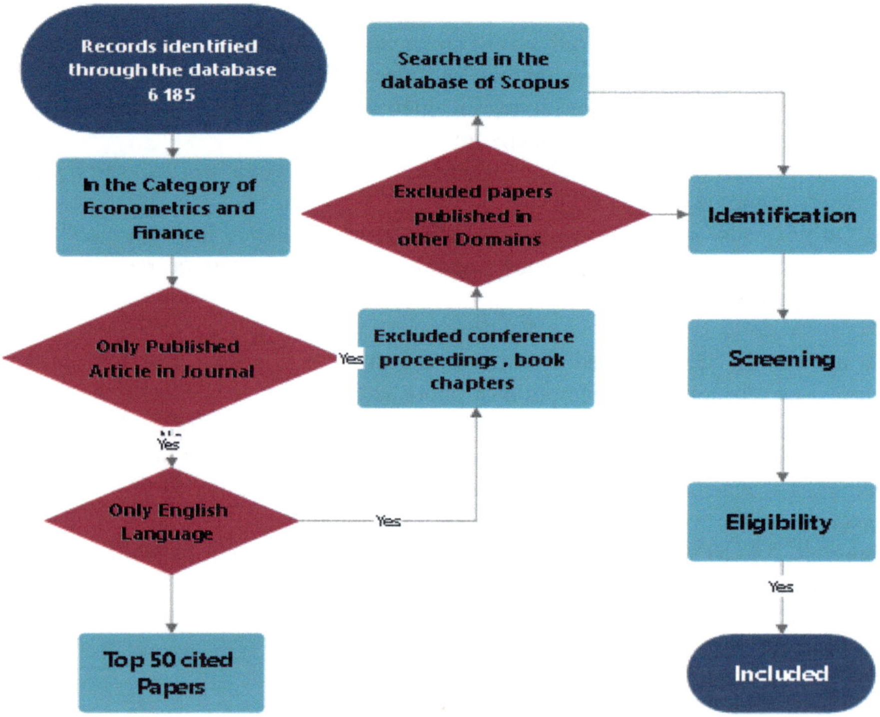

Fig. 5.1 Flow chart of review methodology. *Source* Authors' Own

5.4 Theories Used in Papers

5.4.1 Efficient Market Hypothesis (EMH)

A theory of finance known as Efficient Market Hypothesis (EMH) contends that financial market endlessly resonances all information that is accessible. Financial markets can be understood using the central impression offered by the efficient market hypothesis (Pereira et al. 2022; Ante 2022). Put contrarily, the theory proclaims that asset value precisely signifies all accessible data and promptly adjust to novel information at it becomes accessible. According to the Efficient Market Hypothesis (EMH), asset prices react quickly to new information, meaning that all of the data regarding to asset has already been factored in. This would imply that in the case of NFTs, the market price of that NFT should quickly adapt to reflection of new information that becomes available. Speculative bubbles can happen, despite the Efficient Market Hypothesis (EMH) signifying that markets are efficient (Özdemir et al. 2022; Zhang et al. 2023). This means that assets, including NFTs, may be determined by fervour or irrational optimism rather than by their true value or by the efficient use of information. Short-term price swings resulting from these circumstances may, in the short run, contradict the EMH (Cogo 2022; Anselmi and Petrella 2023). Some opponents contend that because of asset class's novelty, the subjective assessment of digital information, and the impact of speculative trading, nascent markets such as NFTs are not entirely efficient (Aboobaker 2023). These reasons suggest that NFT markets, particularly in the early stages of market growth, may show features more in line with semi-strong or even weak form inefficiencies.

5.4.2 TAM Theory

Presented by Fred Davis in the year 1986, Technology Acceptance Model (TAM) acts as a theoretic structure envisioned to clarify, explain, and investigate the acceptance of technology by the end users. This framework has expanded broadly and has been accredited across diverse subdivisions to discover the procedure and progression through which individuals hold and apply new and updated technical developments. The application of this framework spans across plentiful domains such as information systems, advertising, promotion, healthcare, education, amongst others. By extending support to the investigators and practitioners, this model sheds light on the fundamentals of technology acceptance and devices approaches to enlarge both receipt and application (Perez et al. 2023; Toraman and Geçit 2023; García-Monleon et al. 2022). An example for understanding how individuals come to hold and apply the technology, make use of any new technology is termed as the Technology Acceptance Model (TAM). It's particularly relevant while discussing about emerging ideas like Non-Fungible Tokens (NFTs), since it narrates how consumers will accept and utilize them (Kılıçaslan and Ekizler 2022).

The TAM model offers a prearranged technique for examining the variables affecting adoption of NFT. It suggests that for NFTs to emerge as widely recognized technology, it is important that users must perceive NFTs as practical and modest to use. This thought will definitely inspire positive attitudes and behavioural intentions (Nevi 2022).

5.4.3 Unified Theory of Acceptance and Use of Technology (UTAUT)

Mounting upon and integrating with other pre-existing frameworks, such as TAM, UTAUT is a vigorous theoretic contextual framework that broadens people's understanding of, in what manner individuals make use of technology and accept technology. Through the review of abundant variable quantity and their association, UTAUT offers valued viewpoints on tactics for nurturing technology acceptance and competent, well-organized use across varied fields (Ho and Song 2023). UTAUT has been broadly used to review the adoption and use of the knowledge in a variety of contexts, such as consumer behaviour, healthcare, education, organizational settings, and more. When using the UTAUT model to inspect NFTs, it's evident that several elements, in accumulation to just how they valued and easy to use, are important in defining their adoption. The effect of social factors, the obtainability of resources that make use easier, and the overall experience with NFT platforms and dealings are all substantial in moulding people's opinions and actions regarding the acceptance and use of NFTs (Cho 2022). This feature looks at how social essentials affect people's inclination to trial a technology. For NFTs, social fundamentals could be from fiends, influencers, or groups within the blockchain and digital art communities. Positive endorsements from influential people can motivate adoption, whereas negative views from reliable sources can detect it (Lee 2022; Ramly 2024; Vega 2024). This is about the readiness of an operator to tryout a specific behaviour—in this situation, using NFTs. Essentials such as the belief in technology's effectiveness, the effort required, the impact of social factors, and the accessibility of resources all play a role in shaping people's intent to adopt NFTs.

5.4.4 Modern Portfolio Theory (MPT)

Harry Markowitz created the Modern Portfolio Theory (MPT), a outline for construction an ideal portfolio by either lessening risk for a assumed level of return or maximizing returns for a assumed level of risk. Depositors can lessen risk by distributing their holdings across a variety of NFTs (art, collectibles, virtual real estate, etc.), just like they can with traditional assets. Additionally, because NFTs are exceptional and regularly liquid, defining the expected returns and risks related with them can be more problematic (Lommers et al. 2023; Hemrit et al. 2023).

However, analysing issues such as past price trends, the creator's reputation, and market demand can benefit.

5.5 Methodologies Explain in the Manuscripts

5.5.1 Data Collection Platforms

The study is highlighting a wide range of platforms utilized for data collection related to financial markets and digital assets like cryptocurrencies, NFTs, other traditional assets. Data collection for analytical purposes predominantly relies on various online platforms and sources. Notably, studies frequently cite nonfungible.com (e.g. Aharon and Demir 2022; Zhang et al. 2022; Umar et al. 2022a, b, c; Wang 2022), www.coingecko.com (e.g. Qiao et al. 2023), www.investing.com (e.g. Zhang et al. 2022; Ghosh et al. 2023a, b), and https://coinmarketcap.com/ (e.g., Chowdhury et al. 2023; Ghosh et al. 2023a, b), Yousaf and Yarovaya 2022a, b; 2023; Wang 2022). Additionally, researchers also utilize traditional financial data sources such as Bloomberg and DataStream, Thomson Reuters (Like Goodell et al. 2023; Chowdhury et al. 2023, Umar et al. 2022a, b, c). The diversity reflects the need to access specific types of data relevant to different research domain.

5.5.2 Variables Identified

Most of the previous studies conducted by the scholars, academicians focused on investigating a varied range of assets across numerous categories. Cryptocurrencies have garnered noteworthy consideration, predominantly Bitcoin (Dowling 2022a, b; Umar et al. 2025; Khati et al. 2022; Wang 2022; Yousaf et al. 2023; Ghosh et al. 2023a, b; Zhang et al. 2023), and Ethereum (Aharon et al. 2024; Ghosh et al. 2023a, b; Zhang et al. 2023) as their price movements and market behaviour are closely inspected in experimental analyses. Together with these cryptos, non-fungible tokens (NFTs) have arose as a distinctive group of digital assets, with studies often featuring prevalent tokens such as Decentraland LAND (Dowling 2022a, b), CryptoPunks (Wang 2022) and assets from platforms like Axie Infinity (Dowling 2022a, b). List of key variables has been presented in the Table 5.1.

In the arena of decentralized finance (DeFi), academicians repeatedly discover assets like AAVE, BANCOR, LINK, and MKR (Ghosh et al. 2023a, b) which are vital in understanding the emerging landscape of block chain-founded financial services. These belongings permit activities like yield farming and liquidity provision across decentralized platforms, pinpointing on-going innovations in the domain of financial technology.

Old and outdated financial tools continue to be vital as well, with studies referencing bullions, equities, debts, major currency pairs (F.X.), crude besides other plentiful financial indices like MSCI World Equity and FTSE World Government Bond indices (Aharon et al. 2024; Umar et al. 2025; Xia et al. 2022; Zhang et al.

Table 5.1 List of variables used in the previous studies

Variable	Types
Cryptocurrencies:	• Bitcoin • Ethereum • Binance coin (BNB) • XRP (ripple) • ADA (Cardano) • Cryptocurrency coins • Bitcoin, ethereum, binance coin, and XRP
Non-fungible tokens (NFTs):	• Decentraland LAND tokens • CryptoPunk images • Axie infinity game characters • THETA • Tezos (XTZ) • Enjin coin (ENJ) • NFTs (art, collectibles, games, metaverse, utilities) • (Mana, theta, enj, and waxp)
Decentralized finance (DeFi) assets:	• AAVE • BANCOR • LINK • MKR (maker) • yield farming tokens • DeFi tokens • GameFi, SciFi, SocialFi, and others metaverse-based jobs, such as metaverse-based academies and markets
Financial markets	• Equities • Bonds • Currencies (F.X.) • Gold • Oil • MSCI world equity index • FTSE world government bond index • commodity index

2022). These possessions and assets offer benchmarks against which the performance of cryptocurrencies namely NFTs, and DeFi assets can be calculated and examined.

Overall, the blend of cryptocurrencies, NFTs, DeFi assets, and outmoded financial tools signifies a comprehensive methodology in modern research, directing to unravel the interconnected dynamics and implications of these diverse asset classes in international financial marketplaces and technical progressions.

5.5.3 Techniques Data Analysis

The current research work investigated the data analysis techniques adopted and used in the previous research works and investigations. Through in-depth analysis, it was found that many studies have used and applied the technique known as Time-Varying Parameter Vector Autoregressive (TVP-VAR) model for investigating the dynamic relations amongst economic variables. This technique licenses flexibility in apprehending changes in parameters and has been used to evaluate impact of economic policies, financial market dynamics, and macroeconomic prediction as used by Dowling (2022a, b), Aharon et al. (2022), Wang (2022), Umar et al. (2022a, b, c), Goodell et al. (2022) along with others.

- **TVP-VAR Method**: Abundant studies conducted in the past time utilized the Time-Varying Parameter Vector Autoregressive (TVP-VAR) model for investigating dynamic relations between dissimilar financial variables over the period.
- **Volatility Spillover (Diebold and Yilmaz** 2012): Many previous researches have adopted this methodological framework for evaluating how shocks, instability, and unpredictability in one marketplace interrupt others marketplaces. Studies have applied this approach to assess interconnectedness and systemic risk in fiscal markets, identifying channels through which volatility, instability, and unpredictability spreads across assets and regions.
- **Wavelet Coherence**: This technique of data analysis was adopted in studies intent on understanding the harmonization and relationships amid time series data at different occurrences.
- **NARDL Model**: Nonlinear Autoregressive Distributed Lag (NARDL) models have been recognized to study asymmetric impact and long-run relations in numerous economic variables. These techniques are useful in apprehending nonlinearities and irregularities in financial and economic data, making them appropriate for investigating complex economic phenomena.
- **Structural Equation Modelling (SEM)**: SEM is extensively utilized in social sciences and economics studies for investigating dynamic and complex relationships amid multiple variables. Investigators and academicians apply SEM to examine theoretical models, validating hypotheses, and understanding fundamental and casual relations amid variables.

5.6 Context in the Selected Papers

The previous research works involved in the systematic literature review (SLR) outlined numerous important objectives related with crypto currencies and NFTs. This includes objectives like review of pricing dynamics (Dowling 2022a, b), testing market efficiency (Chowdhury et al. 2023), and the detection of possibilities for arbitrage (Zhang et al. 2022) or marketplace abnormalities. Besides, the work revealed the interconnectedness among crypto currencies/NFTs and conventional financial markets (Aharon and Demir 2022; Umar et al. 2022a, b, c; Yousaf and Yarovaya 2022a, b), studying spillover effects and relations with other asset classes. They also observe patterns of volatility (Wang 2022), causal factors, and measures of risk management like hedging possibilities and volatility spillovers

Table 5.2 Explanation of concepts used in papers

Domain	Description
Price and market efficiency	Describe and analyse pricing dynamics of crypto currencies/NFTs. Assess market efficiency or inefficiency. Identify arbitrage opportunities or market anomalies
Connectedness and integration	Investigate interconnectedness between crypto currencies/NFTs and traditional financial markets. Analyse spill over effects and correlations with other asset classes
Risk and volatility	Examine volatility patterns and factors influencing volatility. Assess risk management strategies and hedging opportunities. Explore volatility spillovers
Investor behaviour and sentiment	Study investor sentiment and its impact on asset prices. Analyse trading volumes and patterns (e.g. herding behaviour, quantile connectedness). Investigate whether NFTs can act as a hedge and safe against traditional assets
Economic and industry impacts	Evaluate economic implications of cryptocurrencies/NFTs on industries. Assess the value proposition of NFTs in creative industries and beyond
Technological advancements and innovation	Investigate block chain technology advancements and applications. Assess the role of DeFi in the crypto currency ecosystem. Study innovations in exchanges
Long-term trends and future outlook	Predict future trends in cryptocurrency/NFT markets. Discuss long-term investment strategies and portfolio diversification. Explore potential evolution

(Goodell et al. 2023). Investor behaviour and mood are examined, with a prominence on their impact on asset prices, trading volumes, and patterns, as well as the potential for NFTs to operate as hedges against traditional assets. Furthermore, the economic and industry consequences of crypto currencies and NFTs are analysed, with a focus on the value proposition of NFTs in creative areas. Following table states the domain of the framed objectives along with its description (as per Table 5.2).

5.7 Future Scope and Directions

The previous studies examined in this systematic literature review suggested numerous future paths for research in the realm of crypto currencies and NFTs. These studies recommend investigating the relationship between NFTs and Ethereum using a variety of uncertainty indicators, including the Volatility Index

(VIX), Economic Policy Uncertainty (EPU), Consumer Confidence Index (CCI), and Consumer Sentiment Index (CSI). They also advocate analysing how economic and financial policy uncertainty, as well as volatility indices such as VIX, MOVE, and TYVIX, affects NFT pricing and market sentiment. Another suggestion is to expand the GARCH-MIDAS model to include additional financial markets and determine whether NFT-related attention influences financial market volatility. Furthermore, qualitative research approaches such as phenomenology and ethnography are advocated for studying NFT entrepreneurship, with a focus on NFT entrepreneurs' values, motivations, and social interactions. It is also recommended to compare the links between returns and trading volumes, as well as volatility and trading volumes, for standard cryptocurrencies, NFTs, and DeFi assets. The studies also propose for broadening the scope to cover more digital asset classes for comparison research. Finally, future research should focus on market sentiment in NFT and DeFi markets, as well as establishing novel forecasting approaches for pricing dynamics in these digital asset classes. Table 5.3 contains a full summary of these main research directions.

Table 5.3 Future scope of research

Domain	Description
Connectedness with uncertainty measures	Examining how NFTs and Ethereum are connected to changes in uncertainty metrics such as volatility index (VIX), economic policy uncertainty (EPU), consumer confidence index (CCI), and consumer sentiment index (CSI)
Impact of economic and financial policy uncertainty	Assessing how economic and financial policy uncertainty, as well as pure volatility metrics (e.g. VIX, MOVE, TYVIX), influence NFT pricing and market sentiment
GARCH-MIDAS model extensions	Intensifying the GARCH-MIDAS methods to include other financial markets and assessing whether financial market volatility is driven by attention to NFTs utilizing prediction power evaluation methods
Comparison of return-volume and volatility-volume nexus	Comparing the relationship between returns and trading volumes, as well as volatility and trading volumes, across traditional cryptocurrencies, NFTs, and DeFi assets simultaneously
Inclusion of other digital asset classes	Widening the inquiry by including other digital asset classes such as cryptocurrencies and decentralized financial assets for comparative analysis with NFTs
Market sentiment and forecasting methodologies	Studying market sentiment in NFTs and DeFi markets, developing forecasting methodologies for pricing dynamics in these digital asset classes

5.8 Implications

As a result of the study, gaps in the prevailing and previous research as well as areas that have not yet been meticulously investigated were revealed, which delivers crucial advice for future examinations. Through an in-depth appraisal of the prevailing body of literature, the study has exposed noteworthy voids and uncharted territories in the field of NFT research, therefore indicating unequivocally the requirement for supplementary investigation. The comprehensive documentation of study gaps has presented valuable visions and way for upcoming study, foremost academics towards addressing these subjects that have been unnoticed despite their importance. Therefore, as a significance of this, the field of study on NFT will be meaningfully expanded. In accumulation to this, the study has provided a considerable outline for the building of future research by accurately authenticating the method that was utilized in earlier inquiries. This framework makes it easier to imitate fruitful procedures and also makes it possible to identify limits in existing methods, which eventually leads to the expansion of research designs that are more vigorous and ground-breaking. It is likely for investigators to make use of this context in order to confirm that their studies are procedurally rigorous and that they produce outcomes that are consistent and authentic.

The current appraisal has made a very important contribution to the augmentation of the general individual's understanding of non-fungible tokens (NFTs), their penalties, and their role in the digital economy. Because the review has summarized multi-layered concepts and interrelated them in a way that is easily understood, it has made it easier for a greater viewer to understand the inference of non-financial transactions (NFTs). It is vital to improve this knowledge because it enables operators to contribute in non-fungible tokens (NFTs) in an active and rational manner, irrespective of whether they are in view of making an investment, producing a new token, or merely exploring this escalating topic. The research work highlighted the impact that non-fungible tokens (NFTs) have had on a variety of businesses, including loaning, betting, entertainment, and the arts. The practices that artists use to generate, indorse, and disseminate their artwork within the scope of art have been profoundly modified because of the introduction of NFTs. In this work, we reviewed the many methods in which non-fungible tokens (NFTs) provide artists the opportunity to earn money while also pronouncing their imaginative side. In addition to this, it reviewed the challenges that are connected with the shield of intellectual property rights and the administration of market volatility. NFTs are instigating a revolution in the entertainment sector, particularly in terms of fan engagement and the monetization of content. This is because personalities and producers are using NFTs to give their followers with one-of-a-kind content and experiences.

Non-financial transactions (NFTs) present novel asset potentials and at the same time also pose challenges to the banking sector. In this current research work, the economic, financials, monetary potentials of non-fungible tokens (NFTs) were inspected. Precisely, the current research focused on the application of NFTs as guarantee in decentralized finance (DeFi) claims and their amalgamation into traditional financial institutions. While this was going on, the assessment highlighted the risks that relate to non-financial transactions (NFTs), such as speculations in the marketplace, governing vagueness, and the possibility of fraud and scams. Through the demonstration of an inclusive lookout on the assistances and challenges related with the NFT setting, the appraisal delivers shareholders with the support they need to circumnavigate the multifaceted setting in an effective manner.

Through the application of novel possession and monetization preparations for in-game assets, non-fungible tokens (NFTs) are bringing about a revolution in the gaming industry. During the assessment, the competence of non-fungible tokens (NFTs) to deliver players with complete possession of them in-game assets was inspected. This would enable players to trade or sell them in-game assets outside of the gaming environment, therefore creating actual value in the real world. The current trend of players gaining possession of in-game assets is instigating a essential shift in the betting economy. This change has prompted anxieties about the upcoming of digital asset ownership as well as the prospect that these assets may be utilized in other games (Chalmers et al. 2022).

It is the opinion of specialists that depositors would be better able to make knowledgeable decisions on their contribution in the NFT domain if they obtain these thorough understandings. This is factual irrespective of whether the investors are procurers, vendors, or inventers. A inclusive analysis of the economic opportunities and risks related with non-financial instruments (NFTs) is providing by the research. This inclusive analysis provides investors with a hard substance upon which to assess the chances and challenges that they may face. When investors hold comprehensive understanding of the intricacies of the NFT market, they will be able to improvise their tactics and evade making the mutual blunders. It is crucial to build regulatory outlines that safeguard consumers and motivate unbiased and translucent markets in order to promise the development and firmness of the NFT market. This will safeguard that the industry lasts to enlarge and continue to be stable. The assessment provides a substance for policy representatives to discourse on the utmost effective methods to address the unique tests that are provided by non-financial instruments (NFTs) by providing an examination of the regulatory framework and its potential significances on the NFT market.

By identifying significant themes for supplementary investigation, providing guidance on research methods, refining public understanding, and bestowing a fair valuation of the financial chances and concerns linked to NFTs, the current assessment has made a notable effect on the field of NFT research. In conclusion, the review has made a considerable contribution to the field. This study takes a detailed look at these multifaceted issues, laying the basis for more well-informed and impactful research, policy-making, and investment in the non-financial skill industry.

5.9 Conclusion and Limitations

Specifically, we have identified theoretical inconsistencies and knowledge gaps that future research should resolve regarding the following topics: Examining how NFTs and Ethereum are connected to changes in uncertainty metrics such as Volatility Index, studying market sentiment in NFTs, and DeFi markets, comparing the relationship between returns and trading volumes, expanding the GARCH-MIDAS model to include other financial markets and evaluating whether financial market volatility is driven by attention to NFTs using prediction power evaluation methods, assessing how economic and financial policy uncertainty, as well as pure volatility metrics (e.g. VIX, MOVE, TYVIX), influence NFT pricing and market sentiment. For each of the topics, we provide arguments for why are relevant to close the knowledge gap or to resolve the conflicting theoretical predictions and inconsistencies encountered. The opportunities for future research that we have identified should also spur the debate on whether NFTS are better investment options than traditional financial assets. Using only one database for collecting papers is the major limitation of the study. Future studies can be performed using databases other than Scopus. The time period covered is limited. Future studies can perform longitudinal analysis. The study had reviewed top cited 50 papers, and in the future, more detailed study can be conducted for presenting a more holistic conclusion. Number of citations was considered as a crucial factor for selecting the papers. In the future, other factors can also be considered for selecting the papers.

References

Aboobaker M (2023) The impact of retail investor attention on the returns, trading volume and volatility of non-fungible tokens. http://hdl.handle.net/11427/39178

Aharon DY, Demir E (2022) NFTs and asset class spillovers: lessons from the period around the COVID-19 pandemic. Financ Res Lett 47:102515. https://doi.org/10.1016/j.frl.2021.102515

Aharon DY, Alon I, Vakhromov O (2024) Metaverse tokens or metaverse stocks–Who's the boss? Res Int Bus Fin 69:102259

Anselmi G, Petrella G (2023) Non-fungible token artworks: more crypto than art? Financ Res Lett 51:103473. https://doi.org/10.1016/j.frl.2022.103473

Ante L (2022) The non-fungible token (NFT) market and its relationship with bitcoin and Ethereum. FinTech 1(3):216–224. https://doi.org/10.3390/fintech1030017

Ante L (2023) Non-fungible token (NFT) markets on the Ethereum blockchain: temporal development, cointegration and interrelations. Econ Innov New Technol 32(8):1216–1234

Arcenegui J, Arjona R, Baturone I (2020) Secure management of IoT devices based on blockchain non-fungible tokens and physical unclonable functions. In: Applied cryptography and network security workshops: ACNS 2020 satellite workshops, AIBlock, AIHWS, AIoTS, Cloud S&P, SCI, SecMT, and SiMLA, Rome, Italy, October 19–22, 2020, proceedings, vol 18. Springer international publishing, pp 24–40

Arcenegui J, Arjona R, Román R, Baturone I (2021) Secure combination of IoT and blockchain by physically binding IoT devices to smart non-fungible tokens using PUFs. Sensors 21(9):3119. https://doi.org/10.3390/s21093119

Arora A, Kanisk, Kumar S (2022) Smart contracts and NFTs: non-fungible tokens as a core component of blockchain to be used as collectibles. In: Cyber security and digital forensics: proceedings of ICCSDF 2021. Springer, Singapore, pp 401–422

Arora S, Satija A, Mal P (2025) The modern finance: upcoming financial instruments & their integration in sustainable models for businesses. Effulgence-A Manag J 23(2):41–44. https://doi.org/10.33601/effulgence.rdias/v23/i2/2025/41-44

Bamakan SMH, Nezhadsistani N, Bodaghi O, Qu Q (2022) Patents and intellectual property assets as non-fungible tokens; key technologies and challenges. Sci Rep 12(1):2178. https://doi.org/10.1038/s41598-022-05920-6

Behl A, Pereira V, Nigam A, Wamba S, Sindhwani R (2024) Knowledge development in non-fungible tokens (NFT): a scoping review. J Knowl Manag 28(1):232–267

Bouraga S (2021) On the popularity of non-fungible tokens: preliminary results. In: 2021 3rd Conference on blockchain research & applications for innovative networks and services (BRAINS). IEEE, pp 49–50

Chalkias KK, Kostis A, Alnuaimi A, Knez P, Naulty J, Salmasi A, Servatius R, Veloso R (2024) Preserving Nature's Ledger: Blockchains in Biodiversity Conservation. arXiv preprint. https://arxiv.org/abs/2404.12086. https://doi.org/10.48550/arXiv.2404.12086

Chalmers D, Fisch C, Matthews R, Quinn W, Recker J (2022) Beyond the bubble: will NFTs and digital proof of ownership empower creative industry entrepreneurs? J Bus Ventur Insights 17:e00309

Chandra Y (2022) Non-fungible token-enabled entrepreneurship: a conceptual framework. J Bus Ventur Insights 18:e00323

Chiacchio F, D'Urso D, Oliveri LM, Spitaleri A, Spampinato C, Giordano D (2022) A non-fungible token solution for the track and trace of pharmaceutical supply chain. Appl Sci 12(8):4019

Cho KH, Lee WB (2022) A study on influencing factors on intention to adopt NFTs using UTAUT. J Korea Contents Assoc 22(3):17–34

Chowdhury MAF, Abdullah M, Alam M, Abedin MZ, Shi B (2023) NFTs, DeFi, and other assets efficiency and volatility dynamics: an asymmetric multifractality analysis. Int Rev Financ Anal 87:102642

Cogo V (2022) The irrationality behind cryptocurrencies. http://hdl.handle.net/10579/22371

Cornelius K (2021) Betraying blockchain: accountability, transparency and document standards for non-fungible tokens (NFTs). Information 12(9):358

Cornelius KB (2020) Smart contracts as evidence: trust, records, and the future of decentralized transactions. Second Int Handb Internet Res:627–646

Das D, Bose P, Ruaro N, Kruegel C, Vigna G (2022) Understanding security issues in the NFT ecosystem. In: Proceedings of the 2022 ACM SIGSAC conference on computer and communications security, pp 667–681

Diebold FX, Yilmaz K (2012) Better to give than to receive: predictive directional measurement of volatility spillovers. Int J Forecast 28(1):57–66

Doan AP, Johnson RJ, Rasmussen MW, Snyder CL, Sterling JB, Yeargin DG (2021) NFTs: key US legal considerations for an emerging asset class. J Tax Invest 38(4)

Dos Santos RB, Torrisi NM, Pantoni RP (2021) Third party certification of agri-food supply chain using smart contracts and blockchain tokens. Sensors 21(16):5307

Dowling M (2022a) Fertile LAND: pricing non-fungible tokens. Financ Res Lett 44:102096

Dowling M (2022b) Is non-fungible token pricing driven by cryptocurrencies? Financ Res Lett 44:102097

Fairfield JA (2022) Tokenized: the law of non-fungible tokens and unique digital property. Ind LJ 97:1261

Far SB, Bamakan SMH, Qu Q, Jiang Q (2022) A review of non-fungible tokens applications in the real-world and metaverse. Procedia Comput Sci 214:755–762

Far SB, Rad AI, Asaar MR (2023) Blockchain and its derived technologies shape the future generation of digital businesses: a focus on decentralized finance and the Metaverse. Data Sci Manag 6(3):183–197

Ferone A, Della Porta A (2022) A blockchain-based infection tracing and notification system by non-fungible tokens. Comput Commun 192:66–74. https://doi.org/10.1016/j.comcom.2022.05.027

García-Monleon F, Erdmann A, Llorente RA, Hernández RF (2022) Adopción de NFT como arte digital: El efecto moderador de la cripto-experiencia y el papel de la sostenibilidad. TECHNO REVIEW Int Technol, Sci Soc Rev/Rev Int Tecnol, Cienc Y Soc 11(4):1–11

Garnett AG (2022) NFTs and the environment. Retrieved (on 30/11/2022) from: https://www.Investopedia. Com/nfts-and-the-environment-5220221# citation-16

Ghosh B, Bouri E, Wee JB, Zulfiqar N (2023a) Return and volatility properties: stylized facts from the universe of cryptocurrencies and NFTs. Res Int Bus Financ 65:101945. https://doi.org/10.1016/j.ribaf.2023.101945

Ghosh I, Alfaro-Cortés E, Gámez M, García-Rubio N (2023b) Prediction and interpretation of daily NFT and DeFi prices dynamics: inspection through ensemble machine learning & XAI. Int Rev Financ Anal 87:102558

Ghosh B, Gubareva M, Zulfiqar N, Bossman A (2024) Is there a nexus between NFT, DeFi and carbon allowances during extreme events?. China Fin Rev Int 14(3):456–479

Goodell JW, Yadav MP, Ruan J, Abedin MZ, Malhotra N (2023) Traditional assets, digital assets and renewable energy: investigating connectedness during COVID-19 and the Russia-Ukraine war. Financ Res Lett 58:104323

Hemrit W, Benlagha N, Ben Arous R, Ben Arab M (2023) Exploring the time-frequency connectedness among non-fungible tokens and developed stock markets. Intell Syst Account, Financ Manag 30(4):192–207

Ho RC, Song BL (2023) User acceptance towards non-fungible token (NFT) as the FinTech for investment management in the metaverse. In: Strategies and opportunities for technology in the metaverse world. IGI Global, pp 59–77

Hufnagel S, King C (2023) Non-fungible tokens: art and crime in a virtual world

Jayasuriya D, Sims A (2023) Not so new kid on the block: accounting and valuation aspects of non-fungible tokens (NFTs). J Risk Financ Manag 16(11):465

Khati P, Shrestha AK, Vassileva J (2022) Non-fungible tokens applications: a systematic mapping review of academic research. In: 2022 IEEE 13th annual information technology, electronics and Mobile communication conference (IEMCON). IEEE, pp 0323–0330

Kılıçaslan F, Ekizler H (2022) Factors effecting purchase intention in blockchain and NFT (non-fungible token) technologies. J Res Bus 7(2):604–623

Kong DR, Lin TC (2021) Alternative investments in the fintech era: the risk and return of non-fungible token (NFT). Available at SSRN 3914085

Kumar S (2022) Strategic management of carbon footprint using carbon collectible nonfungible tokens (NFTS) on blockchain. Acad Strateg Manag J 21:1–9

Lee UK, Kim H (2022) UTAUT in metaverse: an "Ifland" case. J Theor Appl Electron Commer Res 17(2):613–635

Liu Z, Gao H, Lei H, Liu C, Liu C (2023) Overview of non-fungible token eco-regulation, traceability and data right confirmation security. In: International conference on information science, communication and computing. Springer Nature Singapore, Singapore, pp 95–109

Lommers K, Kim J, Baioumy M (2023) Portfolio optimization for heterogeneous assets: the case for NFTs. Available at SSRN 4639259

Lounge TW (2020) Choosing the right blockchain for your NFT. Medium

Mackenzie S, Bērziņa D (2022) NFTs: digital things and their criminal lives. Crime Media Cult 18(4):527–542

Martha GIR, Warnars HLHS, Prabowo H, Hidayanto AN (2023) Semantic literature review on non-fungible token: expansion area of Usage & Trends. In: 2023 international conference on information management and technology (ICIMTech). IEEE, pp 351–356

Mochram RAA, Makawowor CT, Tanujaya KM, Moniaga JV, Jabar BA (2022) Systematic literature review: blockchain security in NFT ownership. In: 2022 International conference on electrical and information technology (IEIT). IEEE, pp 302–306

Mofokeng N, Fatima T (2018) Future tourism trends: utilizing non-fungible tokens to aid wildlife conservation. Afr J Hosp, Tour Leis 7(4):1–20

Mont CG, Persson S, Sánchez CB (2023) Digital tokens for climate action and nature-based solutions: exploration of opportunities and considerations

Musan DI, William J, Gervais A (2020) NFT. Finance leveraging non-fungible tokens. ImpAl CollE Lond, Dep Comput:1–82

Nadini M, Alessandretti L, Di Giacinto F, Martino M, Aiello LM, Baronchelli A (2021) Mapping the NFT revolution: market trends, trade networks, and visual features. Sci Rep 11(1):20902

Nevi G (2022) NFT–non-fungible tokens how react the consumers? An exploratory analysis for a future integrated acceptance model. In: 21st International marketing trends conference Rome. Sapienza University of Rome, Rome

Nobanee H, Ellili NOD (2023) Non-fungible tokens (NFTs): a bibliometric and systematic review, current streams, developments, and directions for future research. Int Rev Econ Financ 84:460–473

Nogueira A, Marques CG, Manso A, Almeida P (2023) NFTs and the danger of loss. Heritage 6(7):5410–5423

Omar AS, Basir O (2020) Secure anti-counterfeiting pharmaceuticals supply chain system using composable non-fungible tokens. In: Blockchain for cybersecurity and privacy. CRC Press, pp 243–259

Özdemir O, Kayhan F (2022) Lockdown effects of Covid-19 outbreak on market efficiency: evidence from non-fungible tokens. Available at SSRN 4822067

Pereira É, Ferreira P, Quintino D (2022) Non-fungible tokens (NFTs) and cryptocurrencies: efficiency and comovements. FinTech 1(4):310–317

Perez WDD, Prasetyo YT, Cahigas MML, Persada SF, Young MN, Nadlifatin R (2023) Factors influencing non-fungible tokens (NFT) game engagement during the COVID-19 pandemic: the theory of planned behavior (TPB) and hedonic motivation system adoption model (HMSAM) approach. Heliyon 9(9)

Popescu AD (2021) Non-fungible tokens (nft)–innovation beyond the craze. In: 5th International conference on innovation in business, economics and marketing research, vol 32, pp 26–30

Qiao X, Zhu H, Tang Y, Peng C (2023) Time-frequency extreme risk spillover network of cryptocurrency coins, DeFi tokens and NFTs. Financ Res Lett 51:103489

Rahman J, Jin Y (2023) The control of tax corruption: evidence from nonfungible token market in China. J Money Laund Control

Raman R, Raj BE (2021) The world of nfts (non-fungible tokens): the future of blockchain and asset ownership. In: Enabling blockchain technology for secure networking and communications. IGI Global, pp 89–108

Ramly FB, Md Zabri MZ (2024) Unveiling the digital desire: UTAUT analysis of NFT investment intentions in Malaysia. China Financ Rev Int

Rehman W, Zainab H, Imran J, Bawany NZ (2021) NFTs: applications and challenges. In: 2021 22nd International Arab conference on information technology (ACIT). IEEE, pp 1–7

Sabharwal S, Mahalwala R (2025) Decoding the volatility dynamics of cryptocurrency market in India. Effulgence-A Manag J 23(2):45–55. https://doi.org/10.33601/effulgence.rdias/v23/i2/2025/45-55

Sadiqin A (2023) Analysis of NFT (non-fungible tokens) in relation to money laundering. Int J Econ, Manag Account (IJEMA) 1(3):153–158

Toraman Y, Geçit BB (2023) User acceptance of metaverse: an analysis for e-commerce in the framework of technology acceptance model (TAM). Sosyoekonomi 31(55):85–104

Truby J, Brown RD, Dahdal A, Ibrahim I (2022) Blockchain, climate damage, and death: policy interventions to reduce the carbon emissions, mortality, and net-zero implications of non-fungible tokens and bitcoin. Energy Res Soc Sci 88:102499

Turki M, Cheikhrouhou S, Dammak B, Baklouti M, Mars R, Dhahbi A (2023) NFT-IoT pharma chain: IoT drug traceability system based on blockchain and non-fungible tokens (NFTs). J King Saud Univ-Comput Inf Sci 35(2):527–543

Umar Z, Abrar A, Zaremba A, Teplova T, Vo XV (2022a) The return and volatility connectedness of NFT segments and media coverage: fresh evidence based on news about the COVID-19 pandemic. Financ Res Lett 49:103031. https://doi.org/10.1016/j.frl.2022.103031

Umar Z, Alwahedi W, Zaremba A, Vo XV (2022b) Return and volatility connectedness of the non-fungible tokens segments. J Behav Exp Financ 35:100692. https://doi.org/10.1016/j.jbef.2022.100692

Umar Z, Gubareva M, Teplova T, Tran DK (2022c) Covid-19 impact on NFTs and major asset classes interrelations: insights from the wavelet coherence analysis. Financ Res Lett 47:102725. https://doi.org/10.1016/j.frl.2022.102725

Umar M, Qin M, Su CW (2025) Exploring the hedging performance of non-fungible token: novel evidence from world uncertainty. Res Int Bus Fin 77:102931

Valeonti F, Bikakis A, Terras M, Speed C, Hudson-Smith A, Chalkias K (2021) Crypto collectibles, museum funding and OpenGLAM: challenges, opportunities and the potential of non-fungible tokens (NFTs). Appl Sci 11(21):9931

Vega E, Camarero C (2024) What's behind the jpg? Understanding consumer adoption of non-fungible tokens. Int J Consum Stud 48(2):e13014

Wang Q, Li R, Wang Q, Chen S (2021) Non-fungible token (NFT): overview, evaluation, opportunities and challenges. arXiv preprint. https://arxiv.org/abs/2105.07447. https://doi.org/10.48550/arXiv.2105.07447

Wang Y (2022) Volatility spillovers across NFTs news attention and financial markets. Int Rev Financ Anal 83:102313. https://doi.org/10.1016/j.irfa.2022.102313

Xia Y, Li J, Fu Y (2022) Are non-fungible tokens (NFTs) different asset classes? Evidence from quantile connectedness approach. Financ Res Lett 49:103156

Yang YJ, Wang JL (2023) Non-fungible token (NFT) games: a literature review. In: 2023 International conference on cyber management and engineering (CyMaEn). IEEE, pp 251–254

Yousaf I, Yarovaya L (2022a) Herding behavior in conventional cryptocurrency market, non-fungible tokens, and DeFi assets. Financ Res Lett 50:103299

Yousaf I, Yarovaya L (2022b) The relationship between trading volume, volatility and returns of non-fungible tokens: evidence from a quantile approach. Financ Res Lett 50:103175

Yousaf I, Gubareva M, Teplova T (2023) Connectedness of non-fungible tokens and conventional cryptocurrencies with metals. N Amn J Econ Financ 68:101995. https://doi.org/10.1016/j.najef.2023.101995

Zhang Y, Chan S, Chu J, Shih SH (2023) The adaptive market hypothesis of decentralized finance (DeFi). Appl Econ 55(42):4975–4989. https://doi.org/10.1142/9781800615212_0008

Zhang Z, Sun Q, Ma Y (2022) The hedge and safe haven properties of non-fungible tokens (NFTs): evidence from the nonlinear autoregressive distributed lag (NARDL) model. Financ Res Lett 50:103315. https://doi.org/10.1016/j.frl.2022.103315

Dr. Abha Gupta is currently working as an Associate Professor in the Department of Management Studies, Rukmini Devi Institute of Advanced Studies, affiliated to GGSIPU, Delhi. She holds a doctorate in Behavioral Finance from Indira Gandhi National Open University. Dr. Gupta completed her postgraduate and undergraduate from University of Delhi and is JRF qualified. Her research has been published in nationally recognized journals, including Scopus-indexed, ABDC-listed, and UGC-CARE-listed publications. She also authored book chapters published by reputed international publishers such as Taylor & Francis and Emerald. Additionally, contributed curriculum chapters for the Department of Distance and Continuing Education at the University of Delhi.

Dr. Kamini Rai is currently working as an Associate Professor in the Department of Management of RDIAS. She has work experience of 13 years in academics. She regularly contributes to various seminars and conferences both at the national and international level. Dr. Kamini has conducted various workshops and FDPs on Advance Research in Finance in the capacity of co-convener and as a resource person. Her research has been published in nationally recognized journals, including Scopus-indexed, ABDC-listed, and UGC-CARE-listed publications. She also authored book chapters published by reputed international publishers such as Taylor & Francis and Emerald.

Prof. Manoj Kumar Mishra is currently working as an Assistant Professor in the Faculty of Management at Marwadi University, Rajkot Gujarat. He has 15 years of academic experience. He submitted his PhD at Jagannath University Jaipur. He did his MBA from VTU with a specialization in marketing and graduated with a BTech (Mechanical) from AKTU.. His more than twenty-seven papers are published in various National Repute journals, including Scopus indexed (Q1, Q2), ABDC and UGC –Care Listed Journals. Mr. Mishra has conducted more than fifty workshops and FDPs.

Part III

Digital Markets and Behavior

Building a Theoretical Framework of Web-Based Instruction in the Context of Online Education

6

Sorabh Bajaj and Mamta Sharma

Abstract

Through the lens of transactional distance theory (TDT), the web-based instruction (WBI) has been examined for better understanding of online teaching and learning. Various studies have examined the impact of design of WBI on student outcomes. The current research has been explored and mentioned three key design aspects. Firstly, its capacity to deliver content that can be seamlessly revised over a period of time. Secondly, WBI's use of adaptive components that accommodate various learning preferences which allow students to progress with their individual pace. Lastly, 'visual layout' which enables structuring the information on screen and provides smooth navigation also improves the clarity to minimize confusion for learners. Various opportunities need to be opened for learners to engage with the content as well as to receive the concepts clarification and guidance by the instructors. Equally, peer collaboration is crucial among learners through group assignments and other shared projects, which deepens academic comprehension. Moreover, informal communication is one of the dimensions which can foster a stronger feeling to get more connected with respect to social presence. Altogether, this dynamic interplay of dialogue synced with transactional distance theory which emphasized how interaction reduces the sense of separation among participants. Collaboration and autonomy are central to the WBI environment and help learners to decide their goals and organize schedules which further encourages independence, self-directed thinking and builds confidence. Parallelly, cooperation allows them to connect and reflect into a deeper way. The current research highlights WBI working to

S. Bajaj (✉)
Fortune Institute of International Business, New Delhi, India
e-mail: sorabh.bajaj@fiib.edu.in

M. Sharma
Prestige Institute of Management and Research, Gwalior, India

B. Kaur et al. (eds.), *Integrating Industry 5.0 and 4.0 for Technology and Humanity*,
Contributions to Environmental Sciences & Innovative Business Technology,
https://doi.org/10.1007/978-981-92-1780-9_6

ensure the student is engaged and makes a more meaningful learning experience. The WBI approach provides the path to achieve success among learners in their own learning context.

Keywords

Web-based instruction • Online learning framework • Web-based learning • Transactional distance theory • Online learning

6.1 Introduction

In today's era, the internet has completely transformed web applications related to education and instructions. Therefore, web-based instruction (WBI) has become a key focus among academicians as investigated by numerous scholars, specifically focused on the technical aspects of these systems and design guidelines to implement online learning platforms (Ally 2022). However, some researchers faced the multiple challenges also to explicitly connect WBI for teaching and learning, particularly within the context of distance and online education (Garrison 2011). The current study addresses this particular gap by placing WBI widely within online education theory, which aims to improve existing theoretical models after reviewing the current investigations so the gap can be filled with teaching and learning aspects conducted in web-based settings (Anderson and Dron 2011). As mentioned by Garrison and Vaughan (2022), WBI should not be seen automatically as a complete break from earlier pedagogical approaches.

To develop the "understanding about WBI", the current study utilizes several key theories from a teaching perspective, including a few theories, i.e., constructivism and information processing and cognitive flexibility theories, explained 'how learners interact with complex information and develop a way to understand online learning settings (Schneider and Shiffrin 1977; Spiro et al. 1988). WBI is not a completely new form of learning even it is a modern way to deliver education remotely, also belongs to a larger group of technology-based approaches like multimedia, hypermedia, and computer-based instruction, and has further taken many forms from past centuries such as videoconferencing and correspondence courses (Bates 2015; Ohle-Peters 2023; Clark 1983; Jacobson and Spiro 1995). WBI is different because of its ability to connect text, images, sound, and video in one space and links learners to wide information sources and allows flexible, often asynchronous, communication between students and instructors (Anderson 2023). Web-based learning enables students more control toward their study as they can modify their own pace and allow them to make decisions for arranging and sorting their lesson plans. The flexible environment of WBI simplifies the adjustments for different learners as well as supports in enhancing engagement. Online forums and shared spaces help students to work in collaborations and idea sharing, showing technology becomes a real part of learning, not just a digital addition.

WBI connects the classroom learning in regard to daily digital life such as tools like blogs and podcasts allow the students to follow their interests beyond the syllabus. Additionally, online communities provide them space to explore ideas with flexibility. WBI combines structured lessons with open exploration when effectively planned, furthermore, leads to a deeper connected learning experience because structure does not confine students to fixed or teacher-led tasks. WBI provides accessibility to students which enables the information gathering when they need it most. Even the resources or review lessons can be searched at any time; students can also join discussions as per available schedules, which makes the learning timelier and more meaningful. This allows teachers also to adjust lesson plans and improve future course designs as learning analytics can also be shown where students face problems.

WBI allows how teachers and students work could be modified as it is more than the transfer of content to the web. Teachers act as designers, facilitators, and informed decision-makers. Students take active roles and use feedback to improve their learning, also work together responsibly for progression. In this setting, adaptability and regular communication become essential as continuous enhancements improve the process (teaching and learning) in web-based settings. The current study explored key pedagogical aspects after reviewing recent investigations on teaching and learning for WBI, aimed to create a theoretical framework for WBI based on TDI. The review has been done after studying the papers published in peer-reviewed journals within the domain. Also, this reflects continuous discussions and findings within the research community of this domain. This research also shows "how different work describe the design and use of WBI, with the focus of identifying common teaching feature and patterns". The findings have been depicted into a clear connected framework, shows 'how WBI can raise or lower transactional distance' and explains 'how structure, interaction, and student control work combinedly within online learning.'

6.1.1 Transactional Distance Theory

Transactional distance theory (TDT) emphasized how teaching and learning conducts when teachers and students are separated by place or time. The theory was introduced by Moore in the 1970s, focused on three main elements such as dialogue, structure, and learner autonomy. The first element 'dialogue' defined as meaningful conversations between learners and teachers, can possibly happen through discussion and feedback (Anderson 2023). The second element 'structure' mentions 'how well a course fits for different learners'. For instance, a course with bound structure provides fewer choices to students, but a course with flexibility lets students follow their own goals (Garrison and Cleveland-Innes 2025). Learner autonomy has been defined as an ability to set individual goals and manage their progress accordingly and shapes 'how students learn' (Holec 1981).

TDT highlights the gaps in communication and understanding in distance learning. When structure is fixed and dialogue is low, the distance becomes large. In such cases, students depend more on their own autonomy. Strong dialogue and flexible structure help to reduce the distance and enable students to feel supported and connected (Anderson and Dron 2011). Over the period of time, TDT has grown and adapted with new technology. Literature depicts the balancing of dialogue, structure, and autonomy remains central to effective online learning. Continuous efforts are being put by the scholars to improvise these ideas suitable for modern teaching and digital space (Garrison and Akyol 2013). Modern online learning tools are helping in minimizing transactional as they use real-time feedback, flexible systems, and social learning platforms. A stronger link is being created between structure and dialogue through these tools and align learning more closely as per student's needs alongside immediate help and clarity. For example, personal learning paths, interactive materials, and quick communication. This theory plays a very important role in studying and improving online education.

The key features of web-based instruction (WBI) have been discussed in the next sections of this paper. These features are organized through the framework of transactional distance theory; study also advanced the three key ideas of the theory using new data and insights.

6.2 Pedagogical Features of WBI

This research investigated 58 peer-reviewed articles written in English. Six major journals were mainly considered within the domain of distance and online education; The American Journal of Distance Education; Educational Technology Research and Development; British Journal of Educational Technology; Open Learning, Distance Education; and the Journal of Distance Education (Bates 2015; ICDE 2023). As mentioned in the international council for open and distance education index, these are among the most cited journals and also ensure that the review is based on reliable and influential studies (ICDE 2023). These sources were considered as they offered a wide view using WBI in teaching. The web tools helped organize content and guide interaction as per the reviewing the sources (Ally 2022) also shaped how learning tasks are designed and presented. Additionally, it shows that online learning continues to change with advanced technology and new learner needs (Garrison 2011; Garrison and Vaughan 2022). "Interaction is a key theme in this field." A strong WBI includes active exchange among students, even between students and teachers, and also allows interaction with digital content. Discussion forums and wikis support delayed communication, but the video sessions allow real-time exchange. It is mentioned in the research studies that flexibility and access help create student-centered learning and personal learning paths also helps to make courses more effective. The findings from this literature are viewed through transactional distance theory. This theory highlights structure, dialogue, and learner autonomy in distance learning (Moore 1993). Using this lens shows which design features lower or raise the sense of distance in web-based courses. The theory and the research together give useful guidance for WBI design and practice. The aim is to improve teaching quality and learning results in online settings (Anderson and Dron 2011).

6.2.1 General Findings

The studies from the six journals used many research designs. They explored different parts of web-based instruction. Only a few studies examined what happens in daily teaching and learning in WBI. Most focused on course design, interaction, and learner satisfaction. Some also studied how WBI supports self-reported learning. There was less attention on the actual teaching methods used. Research shows that WBI has several main strengths. It supports student-centered learning and uses rich multimedia materials (Ally 2022; Garrison 2011). It also offers high interactivity and adapts to different types of learners. These strengths come from web tools that make content flexible and support many ways to communicate (Bates 2015). Many WBI writers have backgrounds in technology or specific subjects rather than distance learning. This means there are few studies that compare WBI with older distance learning formats like correspondence or television courses (Moore and Kearsley 1996). Claims that WBI builds stronger learning communities are still not proven by enough comparative research (Anderson and Dron 2011; Garrison and Vaughan 2022).

Studies show that WBI offers flexible learning conditions. It supports learners with different backgrounds and study habits (Mayer 2009; Means et al. 2010). Multimedia and interactive tools also fit project-based and constructivist learning. Students build knowledge through tasks and group work (Jonassen et al. 1999; Laurillard 2013). Many studies focus on easy measures like satisfaction or access. Few examine deeper learning results or mental engagement. More long-term research is needed to see how tools, interaction, and autonomy shape skills and learning over time (Table 6.1).

The current research article presents a model of web-based instruction based on transactional distance theory (Moore 1993). The model appears in Fig. 6.1. It organizes the main features of WBI within this framework. The model helps explain how design choices affect teaching and learning in online education. The framework accepts that modern web-based learning goes further than the original theory. New studies highlight ideas such as social presence and data-based personalization (Garrison 2011). They also note complex patterns of online teamwork. Figure 6.1 combines the main ideas of transactional distance theory with these new insights. It gives a clear view of the main teaching features in current web-based instruction (Bates 2015).

Table 6.1 Classification of articles by methodology and medium

	Number of Articles (Percent)
Methodology	
Experiments (true or quasi)	15 (26%)
Developmental studies	18 (31%)
Evaluation studies	13 (22%)
Idea/position papers	12 (21%)
Total	58 (100%)
Major medium	
WBI in specific	30 (52%)
Computer conferencing	13 (22%)
WBI along with other technologies	15 (26%)
Total	58 (100%)

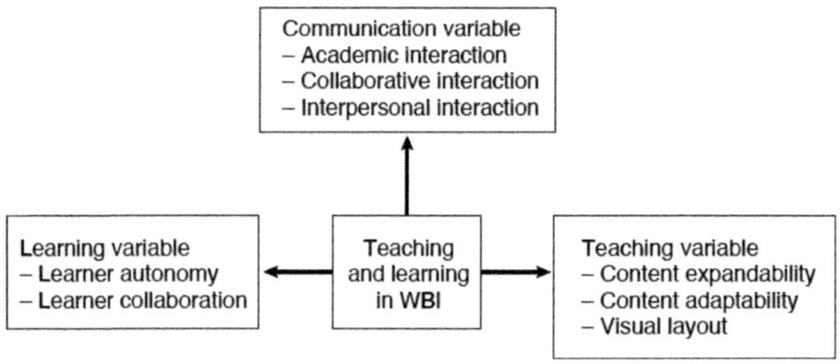

Fig. 6.1 Theoretical framework for pedagogical features of web-based instruction

6.3 Methodology

This study uses a systematic review to explore key teaching elements of web-based instruction (WBI) through transactional distance theory (Moore 1993). The goal is to combine current theory and research about WBI. It studies how courses are structured and how communication happens. It also looks at how learners receive support. These findings are then joined into one clear framework.

The review used four major research databases. These were Scopus, ERIC, JSTOR, and Google Scholar. They were selected because they include a wide range of peer-reviewed studies in educational technology and online learning. The search focused on work published between 2010 and 2025. This ensured attention to current trends in WBI. The search used keywords and Boolean terms such as web-based instruction, online learning, transactional distance theory, online pedagogy, distance learning frameworks, and digital education.

Clear rules were used to choose the studies. Articles had to be peer-reviewed and written in English. They also needed to discuss teaching aspects of WBI. Papers that studied learner interaction, course design, or theory in online education were included. Both research studies and concept papers were accepted. Studies that focused only on technology or system design were removed. Non-academic pieces like blogs or opinion articles were also excluded.

The study selection happened in three steps. First, titles and abstracts were checked to remove unrelated studies. Second, the full papers were read to confirm research quality and relevance. Third, the selected studies were grouped by theme. The main themes were course design, interaction and dialogue, and learner autonomy. This process resulted in 58 papers used for detailed review.

The study used articles from six top journals in distance education and educational technology. These were Educational Technology Research and Development, British Journal of Educational Technology, American Journal of Distance Education, Open Learning, Distance Education, and Journal of Distance Education. The journals were chosen based on their ICDE ranking. This ranking shows their importance and influence in the field (ICDE 2023).

Transactional distance theory (Moore 1993) was used to guide and interpret the studies in this review. Its three main ideas are structure, dialogue, and learner autonomy. These were used to group the findings. This approach gave the study a clear framework while allowing space for new themes that did not fully match the original model.

While reviewing the literature, some studies added new factors or broader meanings to the main ideas of the theory (Garrison and Akyol 2013). These findings were included with the original concepts to show a fuller view of how WBI supports learning. The final framework combines Moore's theory with newer research. It gives a strong, theory-based explanation of teaching and learning in web-based instruction.

6.4 Structural Features in WBI as a Teaching Variable

Research on web-based instruction (WBI) often focuses on communication and interaction. Few studies examine how structured or flexible WBI course designs are (Anderson and Dron 2011; Bates 2015). Most research describes WBI as a flexible form of learning. It connects this flexibility to the use of web technologies (Garrison 2011; Ally 2022).

Most studies agree that WBI supports flexible learning. This flexibility comes from varied content formats and options that meet learner needs (Mayer 2009; Laurillard 2013). Interactive features also help keep students involved. WBI is viewed as well suited for learning spaces that fit different goals and schedules (Anderson and Dron 2011).

Interaction has received more attention than structure in most research. The findings match what is known about online education. Web-based learning uses multimedia tools and lets students control time and pace (Ally 2022). It also supports participation at different times and places. These features make WBI a flexible method of teaching and learning.

More research is needed to understand WBI course structure. Future studies should compare how flexible or fixed WBI designs are when compared to older distance learning models (Bates 2015). This would show how course design affects teaching quality and learning experiences (Garrison 2011).

Studies show that the flexibility of WBI comes from the ability to update and expand content. Courses are not limited to what the teacher prepares. Both teachers and students can add links to trusted sources to build shared knowledge (Laffey et al. 1998). Hyperlinks and guided pathways make it easy to go beyond the main syllabus when needed.

Flexibility in WBI appears in how content fits each learner. A student's goals and knowledge affect what they see and do. Research shows that WBI offers more interaction and response than older distance learning methods (McLoughlin 1999; Valcke and Martens, 1997; Vassileva and Deters 1998). These findings show that WBI can adapt and personalize learning while keeping strong academic standards.

The flexible structure of WBI can also be seen in its screen design. Early research noted that strong web-based courses are easy to use and visually clear (Boshier et al. 1997; McGreal 1997). They also support meaningful interaction. Later studies agree and show that layout and navigation affect how easily students complete tasks and interact with content.

WBI platforms also show flexibility in how they are built. Many have modular designs that let teachers add or change units during the course. This helps them respond to feedback and adjust goals. The modular setup supports different learning paths for different students. Analytics tools make this flexibility stronger by showing progress and engagement. These tools help teachers adjust speed, support, and assessment as the course runs.

Adaptive tools make WBI more flexible for each learner. These systems track student actions and results. They then change the order or level of tasks to fit each learner's needs. Students with different skills or preferences see content in ways that match them better. This kind of change can raise engagement and help students move more smoothly through online learning.

Collaboration tools show how flexible WBI can be. Discussion boards, wikis, and chat spaces let students share ideas and give feedback. They also help learners build knowledge together. Many of these tools work asynchronously, so students can join at any time. This adds time and place flexibility to WBI. Live tools support real-time interaction when quick communication is needed.

Platform design adds flexibility through responsive layouts. Courses that work on laptops, tablets, and phones let students study anywhere. Moving between devices is easy and does not affect how lessons work. This access reduces barriers of time and place. It also helps students keep steady study habits. Adaptive paths, collaboration tools, and responsive layouts show the main strengths of WBI. These include expanding content, adjusting it, and showing it with clear design. These features explain why WBI is seen as a flexible way to teach and learn.

6.5 Dialogue in WBI as a Communication Variable

Meaningful learning in web-based instruction (WBI) involves more than interactions between students and teachers. It also depends on peer relationships and a wider sense of belonging in online learning communities. Early studies linked internet use and perceived benefits to social ties among learners and to their level of social integration. They also showed that interpersonal concerns remain important even in asynchronous online discussions. Scholars note that several kinds of online communities can exist at the same time. One type focuses on academic guidance and expert input. A second centers on peer collaboration and intellectual exchange through discussion and shared tasks. A third highlights emotional support, encouragement, and informal help. Research on collaborative learning supports this view and shows that well-planned computer-mediated communication can sustain effective group work.

Social presence appears again and again in these studies. It plays a major role in how satisfied learners feel and whether they stay engaged in the course. Feeling noticed by others and experiencing a sense of community on the course site are closely linked to ongoing participation. Research also shows that social factors like feeling part of the group and having frequent contact with others increase both actual use of the online environment and the perceived value of learning. In forum posts and live chat, students often greet one another, acknowledge contributions, and use small conversational signals. These actions help them build a shared presence and reinforce community norms.

WBI dialogue works through several channels at the same time. Academic exchanges with teachers and experts support understanding of course content. Collaborative work with peers and informal contact help build trust and motivation. These layers interact and shape how students relate to the material, to one another, and to the course over time.

Research on WBI communication shows that students are more engaged when courses support several kinds of interaction. Academically focused dialogue connects learners with instructors or invited experts and helps clarify concepts. Collaborative interaction occurs when students work together, share ideas, and solve problems in groups. Interpersonal interaction builds social and emotional connection within the class community. Together, these three forms of dialogue show how online courses can be designed to support rich and sustained communication.

These forms of interaction do more than help students grasp the content. They also shape motivation, persistence, and sense of belonging in the course. When learners feel part of a supportive online community, they are more willing to share ideas, ask questions, and admit when they are unsure. This leads to deeper engagement with both the material and their peers.

For instructors, it is important to design courses that include all three types of interaction. The quality and balance of academic, collaborative, and interpersonal dialogue influence how satisfied students feel, how well they perform, and how much of their learning they retain over time.

6.5.1 Learner Autonomy as a Learning Variable

Web-based instruction is often described as a setting where learners can work in ways that match their own goals, pace, and study strategies. Rather than following one fixed route, students can decide when and how they take on tasks, which supports greater autonomy. Earlier work on computer-mediated courses presents these environments as more personalized than traditional, pre-packaged distance materials. Learners are able to plan, monitor, and adjust their own progress in an active way. Studies with adult learners in WBI contexts note frequent use of metacognitive strategies and on-demand searching for information. Many reports show students building their own understanding instead of simply receiving knowledge passively.

Web-based instruction is not limited to solo study. Research shows that students in WBI settings also take part in group tasks and shared activities, often through online discussion spaces that can resemble rich seminar-style exchanges. Comparisons of face-to-face and online conferencing indicate that digital environments can sustain strong student-to-student interaction and collaborative problem solving. Observations of live chat tools further suggest that learners use ongoing questions, clarifications, and feedback to keep a common focus and to lessen feelings of distance from one another.

These findings suggest a dual role for learners in WBI courses. They are expected to take charge of their own progress and also to support the learning of others. In this setting, autonomy does not mean working in isolation. It means making thoughtful choices within a connected network of peers and instructors. Students who succeed in WBI usually build strong self-regulation skills, such as setting goals, managing time, and checking their understanding. They also engage in dialogue so they can test ideas and receive feedback. Well-designed web-based environments help sustain this balance by providing guidance, prompts, and structures that slowly shift more control to learners while still enabling collaboration. This mix of independence and interdependence is a key strength of WBI and fits well with a learner-centered approach to online education.

The digital format of WBI allows learning to be shaped around individual needs instead of a single fixed route. Learners can move through content in ways that match their own goals, using adaptive paths, regular checks for understanding, and feedback based on their activity to see where they are strong and where they need more help. These features give students greater control over how they work with materials and how they move toward their personal learning goals.

WBI can also build strong social and collaborative layers into this personalized structure. Discussion forums, group projects, and shared workspaces give learners space to exchange ideas, question each other's thinking, and build understanding together. This community dimension enriches what might otherwise feel like individual study. When a course intentionally combines learner autonomy with planned opportunities for collaboration, it offers a flexible but reliable framework for growth. Within this framework, students can develop the skills they need to succeed in complex and fast-changing educational and professional settings.

6.6 Discussion

The text argues that more targeted research is needed on the foundations of web-based instruction (WBI), especially on how teaching, communication, and learning variables interact in online settings. Within the teaching or structural dimension, the flexibility of course design is particularly important. How tightly or loosely a course is organized can shape the amount and quality of dialogue, influence transactional distance, and ultimately affect learning outcomes. Flexible designs can support learner autonomy, but they must be planned carefully so that freedom does not result in confusion, disengagement, or weaker performance.

Structural design also includes choices about how content is organized and updated. WBI makes it easier to keep materials current, to divide courses into modules, and to bring text, media, and interactive activities together in one space. These options call for scalable frameworks that can respond to diverse learner needs and contexts. Visual layout is another core structural issue. Strong interfaces offer clear navigation, logical flow, and an appealing look so that learners can focus on understanding the content rather than on working out how to move through the course.

Communication variables are equally central to effective WBI. Different types of interaction between learners and instructors, among peers, and between learners and digital content each play a distinct role in shaping achievement and satisfaction. Designing web-based courses therefore requires a clear understanding of how these interactions provide feedback, enable clarification, and foster a sense of community. Instructors need facilitation strategies that keep some of the conversational feel and responsiveness of face-to-face teaching. Careful use of guiding prompts, timely replies, and well-structured activities supports ongoing engagement even when learners and teachers are separated in time or place.

Differences in cognitive style, confidence with technology, and preferred communication modes influence how they move through online platforms and take part in activities. These differences show why WBI must be designed to be both inclusive and adaptable. Features such as forums, shared documents, and structured peer-feedback tasks are especially helpful when they are purposefully designed to support cooperation and community-building (Ally 2022).

Learning-related variables also require careful attention, especially autonomy and learner empowerment. WBI often gives students considerable control over pace, sequence, and study strategies, but this freedom needs to be balanced with clear guidance and supportive structure (Moore and Kearsley 1996). Too much independence without direction can lead to delay, confusion, or loss of interest.

Effective instructional designs therefore pair choice with scaffolds such as defined milestones, guiding prompts, and timely feedback (Bates 2015; Garrison 2011). Motivation and self-regulation are crucial in this process because they shape how consistently students work toward their goals and how well they perform. These qualities can be strengthened through tools and practices that highlight progress, provide constructive feedback, and use adaptive assessments to adjust challenge levels.

More research is still needed on how particular instructional tools influence understanding and long-term retention. Multimedia resources, interactive simulations, and other rich elements can support learning when they are designed to manage cognitive load rather than overload students. Finding out which blends of tools work best for different kinds of learners will help improve WBI practice. Careful study of teaching design, communication patterns, and learner characteristics can provide a fuller picture of how to optimize web-based instruction both technically and pedagogically.

6.7 Conclusion

This research shows that effective web-based instruction depends on more than using new technologies. It also requires matching those tools with careful, well-thought-out pedagogy. Structural choices such as how tightly a course is organized, how easily content can be adapted, and how clearly information appears on screen all need deliberate planning. The aim is to keep learners engaged and to lower the sense of distance between students and instructors (Moore 1993; Garrison 2011). Flexibility in WBI is a strength, but it must be directed so that students have space for autonomy while still retaining a clear sense of purpose and overall course coherence.

Communication practices are just as crucial as structural design. Different forms of interaction, and especially learner-to-learner exchanges, shape both the social climate of a course and the depth of thinking that develops there (Garrison and Vaughan 2022). Instructors therefore need facilitation approaches that make effective use of online tools to support this interaction. As web-based learning continues to grow, ongoing professional development in online teaching strategies will be essential for maintaining and improving instructional quality (Anderson 2023).

Learners approach web-based instruction with varied ways of thinking, different levels of digital confidence, and differing habits of managing their own study. These differences make it important to design courses that are both flexible and inclusive, offering multiple pathways and interaction formats that can respond to diverse preferences (Ally 2022). Scaffolds such as clear guidance, worked examples, and a gradual release of responsibility help students build autonomy over time. Adaptive technologies can then fine-tune content and support in line with each learner's performance and behavior, while well-designed assessment and feedback systems provide timely, specific information that allows learners to monitor their progress and choose appropriate next steps.

Motivation and engagement are core learning factors that need careful planning. In self-paced and often asynchronous courses, students can find it hard to stay focused or to continue when tasks become challenging (Moore and Kearsley 1996). Designs that include goal-setting activities, chances for peer support, interactive tasks, and varied, dynamic content can help sustain interest and effort. The ability of WBI to personalize resources and activities also makes genuinely learner-centered approaches much more achievable in everyday practice (Bates 2015; Garrison 2011). The use of multimedia and interactive elements also requires careful pedagogical judgment. Visuals, audio, simulations, and other rich media should help to clarify key ideas and manage cognitive load rather than distract or confuse learners. When these tools are closely aligned with clear learning objectives, they can deepen understanding and support longer-term retention of important concepts.

Bringing these strands together-course structure, communication practices, and learner characteristics-helps educators form a clearer theoretical and practical understanding of effective WBI. Transactional distance theory offers a strong foundation by highlighting structure, dialogue, and autonomy as connected levers in course design (Moore 1993). Ongoing research and reflective practice are still needed to refine this framework so that WBI continues to grow as a pedagogically robust, evidence-informed approach that can serve diverse learners fairly and effectively.

References

Ally M (2022) Foundations of educational theory for online learning. In: Anderson T (ed) The theory and practice of online learning. Athabasca University Press, pp 15–44

Anderson T (2023) Getting the mix right again: an updated and theoretical rationale for interaction. Int Rev Res Open Distrib Learn 4(2):9–14

Anderson T, Dron J (2011) Three generations of distance education pedagogy. Int Rev Res Open Distrib Learn 12(3):80–97

Bates AW (2015) Teaching in a digital age: guidelines for designing teaching and learning. BCcampus Open Textbooks

Boshier R, Mohapi M, Moulton G, Qayyum A, Sadownik L, Wilson M (1997) Best and worst dressed web courses: strutting into the 21st century in comfort and style. Distance Educ 18(2):327–348

Bullen M (1998) Participation and critical thinking in online university distance education. J Dist Educ 13(2):1–32

Clark RE (1983) Reconsidering research on learning from media. Rev Educ Res 53(4):445–459

Garrison DR (2011) E-learning in the 21st century: a framework for research and practice. Taylor & Francis

Garrison DR, Akyol Z (2013) The community of inquiry theoretical framework. In: Moore MG (ed) Handbook of distance education. Routledge, pp 104–120

Garrison DR, Vaughan ND (2022) Blended learning in higher education: framework, principles, and guidelines. Wiley

Holec H (1981) Autonomy and Foreign Language Learning. Pergamon

International Council for Open and Distance Education (2023) Open and distance education: global trends in citations. https://www.icde.org/global-trends-citations-report

Jacobson MJ, Spiro RJ (1995) Hypertext learning environments, cognitive flexibility, and the transfer of complex knowledge: an empirical investigation. J Educ Comput Res 12(4):301–333

Jonassen D, Prevish T, Christy D, Stavrulaki E (1999) Learning to solve problems on the web: aggregate planning in a business management course. Distance Educ 20(1):49–63

Laffey J, Tupper T, Musser D, Wedman J (1998) A computer-mediated support system for project-based learning. Educ Technol Res Dev 46(1):73–86

Laurillard D (2013) Teaching as a design science: building pedagogical patterns for learning and technology. Routledge

Mayer RE (2009) Multimedia learning. Cambridge University Press
McGreal R (1997) Information technology and telecommunications: a course on the journal of. Distance Educ 12(1/2):67–84
McLoughlin C (1999) Culturally responsive technology use: developing an online community of learners. Br J Educ Technol 30(3):231–243
Means B, Toyama Y, Murphy R, Bakia M, Jones K (2010) Evaluation of evidence-based practices in online learning: a meta-analysis and review of online learning studies (No. ED505824).U.S. Department of Education
Moore MG (1993) Transactional distance theory. In: Keegan D (ed) Theoretical principles of distance education. Routledge
Moore MG, Kearsley G (1996) Distance education: a systems view. Wadsworth
Ohle-Peters A (2023) Differences in instructional quality during the Covid-19 pandemic: a comprehensive review of web-based instruction. Front Educ 8:1244548. https://doi.org/10.3389/feduc.2023.1244548
Schneider W, Shiffrin RM (1977) Controlled and automatic human information processing: I. Detection, search, and attention. Psychol Rev 84(1):1–66
Spiro RJ, Coulson RL, Feltovich PJ, Anderson DK (1988) Cognitive flexibility theory: advanced knowledge acquisition in ill-structured domains (Technical Report No. 441). University of Illinois at Urbana-Champaign, Center for the Study of Reading
Valcke MMA, Martens RG (1997) An interactive learning and course development environment: context, theoretical and empirical considerations. Distance Educ 18(1):7–23
Vassileva M, Deters R (1998) Dynamic courseware generation on the. Br J Educ Technol 29(1):5–14

Sorabh Bajaj is a forward-thinking AI in Education Leader and Consultant with over 18 years of experience in academia, digital learning, and business strategy. He consults large-scale digital transformation and pedagogical innovation initiatives at Higher Education Institutions in India. he is currently working on leveraging GenAI tools to advance teaching and learning. Sorabh collaborates with faculty to embed AI in curriculum delivery and assessment. He has advised Big 4 consulting firms on addressing training needs for structured problem-solving for their employees. He has presented and published research.

Mamta Sharma is an academician with expertise in teaching and research. She has presented papers at international conferences and actively participates in FDPs. Her interests include operations research and data analysis.

Navigating Challenges, Demands, and Resources in the Gig Economy: A Conceptual Exploration of Well-being among Gig Workers

7

Siao Ying Choo, Mei Peng Low, and Wai Mun Har

Abstract

The paper discusses the issues, needs, and assets of the low-skill physical gig workers, the overlooked group in the findings on the gig economy. It emphasizes the role of these factors affecting their quality of work life, their satisfaction in work, and the quality of their life. The study integrates the job demands-resources model to investigate the implications on whoever working in gig industry. Results indicate that flex and skill are the positive facet of gig work, despite the associated benefits, stress and burnout are other effects of this type of development, as a result of job insecurity, and bad working conditions. The article suggests that better working conditions should be offered by platforms of gig and funds to improve the lives of gig workers. Such limitations, as dependence on, exist secondary data, and the study provides effective information on the impact of issues and resources of gig work toward the quality of the life of the workers. It calls for further empirical studies to build on these findings, especially through surveys in diverse gig environments.

Keywords

Gig economy • Gig work • Challenges • Job demands • Job resources • Work-life quality • Life quality

S. Y. Choo · M. P. Low (✉) · W. M. Har
Universiti Tunku Abdul Rahman (UTAR), Petaling Jaya, Malaysia
e-mail: lowmp@utar.edu.my

S. Y. Choo
e-mail: choosiaoying@1utar.my

W. M. Har
e-mail: harwm@utar.edu.my

B. Kaur et al. (eds.), *Integrating Industry 5.0 and 4.0 for Technology and Humanity*, Contributions to Environmental Sciences & Innovative Business Technology,
https://doi.org/10.1007/978-981-92-1780-9_7

7.1 Introduction

The COVID-19 has risen a remarkable upsurge in individuals engaged in gig work, since they are compelled to remain at home to mitigate the transmission of the virus. As a result, the need for everyday supply and delivery services has increased. According to a projection by Business Research Insights (2024), the worldwide gig industry market priced at USD 0.355 trillion in 2021 and is anticipated to achieve USD 1.864 trillion by 2031. The gig economy in Malaysia has been growing since 2010, but it was not until 2017 that it became widely known (Ghazali and Muhammad 2021). The majority of those working for themselves and residing in urban areas make up the informal sector, according to the International Labor Organization. Harith et al. (2022) discovered that the Klang Valley was home to most low-income families with high unemployment rates. Consequently, by offering work opportunities for this demographic, the gig economy has devoted itself to the reduction of urban poverty and enhancing individuals' wellbeing (Auguste et al. 2022).

Nevertheless, lack of established laws or regulations regarding gig workers' identities makes them vulnerable, particularly with exploitation related to wages and working hours (Loganathan 2022). As a result, many gig workers must put in long hours if they hope to make a living wage. Due to frequent outside labor, gig workers are susceptible to security and health threats. They are not eligible for social security; thus, even with high accident rates, they must cover the expensive cost of medical care out of their own pockets (Kidd 2020). This runs counter to the eighth Sustainable Development Goal (SDG) of the United Nations (UN), which is to safeguard workers' rights and advance a secure workplace for all employees, especially gig workers (United Nations 2020). Fair pay and social security protection are rights and advantages that gig workers ought to be eligible for (Makhtar et al. 2022).

Limited study has examined the causes stimulating the overall contentment of gig workers, particularly those engaged in low-skill physical labor. This research seeks to enhance comprehension of the subject by analyzing the welfare, specifically work-life quality, job satisfaction, and long-term life quality of gig workers committed in low-skill physical labor as the economy serves as a crucial support for the urban impoverished. This emphasis is warranted since low-skill physical laborers are different from high-skill platform workers or traditional employees in their circumstances. This research seeks to explore the results and conclusions from previous research about how these factors affect the welfare of this particular group and then further expand and refine the research gap by focusing on the factors mentioned that are found in the research. Consequently, plans should be developed to improve this set of workers' working conditions.

7.2 Gig Economy

In the early twentieth century, jazz musicians first employed the term "gig" as a term for taped and live performances. Currently, it often denotes a singular musical performance or a provisional role (Cloonan and Williamson, 2023). Huđek and Širec (2023) have presented a concise history of this economy and explored the use of "gig" terminology. Jack Kerouac was the inaugural individual outside the music business to employ the term "gig" to denote a temporary position (Parigi and Ma 2016). Erlich (2021) offers a more pertinent characterization of the "gig economy," defining it by distinct category of flexible work that has expanded after the last recession. According to Hook, as cited in Levitt (2021), the phrase "gig economy" initially described the practice of engaging in many part-time occupations for survival. Meanwhile, Tina Brown, who had previously served as an editor at the New Yorker, came up with the phrase "gig economy" back in 2009 in her articles (Hasija et al. 2020). The significance of information and its influence on the economy is becoming increasingly evident in the increasing prevalence of digital platforms. This is particularly true about the gig economy (Zutshi and Grilo, 2019). Within the gig economy, platform firms use this concept by utilizing data from their applications to connect individual requests and offerings, hence establishing novel business models based on platforms.

Lepanjuuri et al. (2018) stated that this economy is defined by the adaptable and task-oriented exchange of labor for remuneration through digital platforms between people or businesses. The OECD journal characterizes the gig economy as a system of platforms that randomly associate clients with employees on the opposite side (Schwellnus et al. 2019). Some assert that a broader definition of the sharing economy optimizes asset use (Dazzi 2019). A growing belief that the name "sharing economy" is improper for what appears to be a commercial business has emerged (Slee 2017). The idea can no longer define the gig economy, claims Oei (2018). The term "gig economy" denotes an emerging business model characterized by a dual-sided market facilitated by internet expansion. It enables independent contractors to rent out goods or do compensated work (Koutsimpogiorgos et al. 2020).

"Gig economy" denotes a phase of economic transformation when work across multiple sectors is growing increasingly irregular, unstable, and dispersed. Employees are required to work more than one job, even work concurrently, put in less time at one job, assume the risk of temporary unemployment, and spend unpaid time seeking other employment (Woodcock and Graham 2019). Additionally, it discusses job markets where independent contracting takes place on digital media platforms. The kind of work available varies and comprises ad hoc and temporary roles. It can demand piecework compensation, have unpredictable hours and no job security, and offer no chances for career progression.

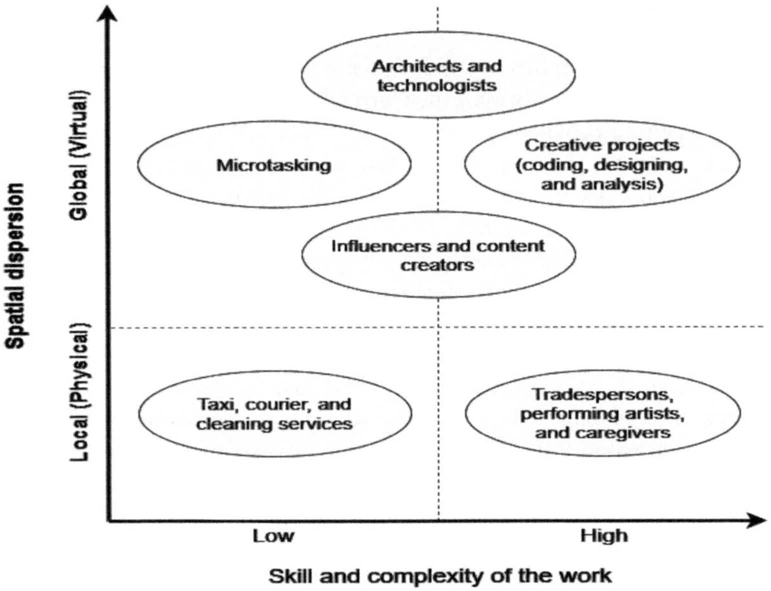

Fig. 7.1 Type of gig work. *Source* Vallas and Schor (2020)

7.3 Classification of Gig Work

Within the gig economy, the crew are regarded by various terms, such as clients, coworkers, freelancers, independent contractors, gig workers, and microbusiness owners (Ahsan 2020). Vallas and Schor (2020) have differentiated across distinct categories of platform work. All the categories are presented in Fig. 7.1 better to comprehend different job types within the economy. These laborers confront instability in the job market and frequently look forward to new opportunities. The skill set that performs the service ranges from low to high skill (top to bottom), and the service delivery mechanism covers both locally (physically) and internationally (virtually). The current study will concentrate more on the third category, which is gig workers with physical and low skill levels.

7.4 Methodology

This section covers the method used for the current research paper. According to Page et al. (2021), A collection of guidelines known as Preferred Reporting Items for Systematic Reviews and Meta-Analyses (PRISMA) was set up to enhance the quality with transparency of current study through an organized reporting framework. The mentioned method (Page et al. 2021) adapted to select materials focusing on gig work, especially low-skill physical tasks. Early studies referred to this work as "crowd work" or "gig economy." Search terms included gig work, gig economy, crowd work, and similar phrases. This provided 523 documents and these were filtered in terms of journal quality, relevance, and additional screening, leaving out irrelevant studies. Of these, 43 studies remained, focusing on issues regulating the well-being of gig workers. In compliance with the PRISMA criteria outlined by Rethlefsen and Page (2022), the abstracts are reviewed to refine the research focus and preserve data that elucidate the determinants influencing the overall contentment of gig workers. Articles that mentioned "gig work" or low-skill work done entirely physically. This kind of employment is distinct from that done remotely, such as Upwork, Task Rabbit, or Freelancer.com, which connects customers and suppliers of remote services. Importantly, PRISMA not only supported the systematic filtering of sources but also informed the structure for the subsequent content analysis, including theme identification and synthesis, to ensure transparency from selection to interpretation. Figure 7.2 is attached to convey the process of article selection better.

7.4.1 Research Setting

A systematic methodology is employed in a systematic literature review (SLR) to gather and assess many reviewed articles or publications. It serves as a comprehensive summary of the body of knowledge pertinent to a certain research inquiry (Mohamed Shaffril et al. 2021). This study employed the SLR procedure with the PRISMA framework as delineated by Page et al. (2021), to effectively tackle the research problem. This methodically done review is classified as an SLR in terms of its kind. The review examines the concept of low-skill physical freelance laborers within the gig economy sector. The subsequent section contains a detailed account of the criteria considered during the development of the literature corpus. The descriptive analysis offers a perceptive synopsis of literature. A component of the approach involves a thematic content analysis of the corpus. This research primarily contributes by arrangement of a concise summary of the ongoing case of knowledge concerning the variables influencing the work-life quality, job satisfaction, and long-term life quality of gig workers.

7.4.2 Sample Selection

There were two stages to the sample selection procedure, and 43 articles were selected as the final sample. Databases, keywords, fields, source kinds, publication stages, search duration, and language were all determined in Phase 1 of the process. The databases accessed for this paper are Science Direct, Emerald Group Publishing, IEEE Xplore, Sage Journals, Scopus, JSTOR, Elsevier, and ResearchGate. The search was temporally unconstrained but restricted to academic journals and English-language publications. After appropriate removals, the initial results of 523 documents were narrowed to 283 documents. The rest of the 240 papers was removed because they did not discuss the topic that the researcher is looking

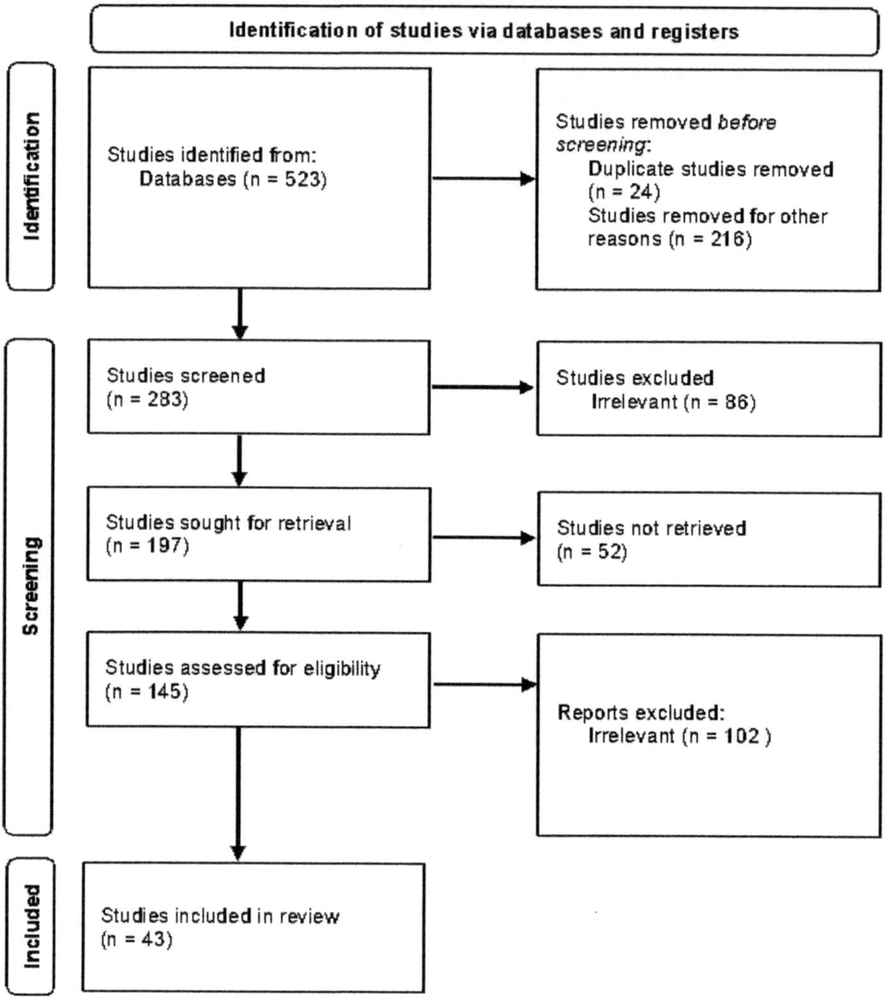

Fig. 7.2 PRISMA flow diagram of this study. *Source* Developed for this study

for within the gig economy context. A quality criterion was used in Phase 2, with an emphasis on peer-reviewed journals with rankings in A, B, or C (VHB) or Q1 or Q2 (SJR). Following screening and quality evaluation, 86 articles were found to be irrelevant; 52 of these were unable to be retrieved, and 102 more were eliminated based on their content relevancy. All 43 of the articles that made up the final selection have been listed in Appendix A.

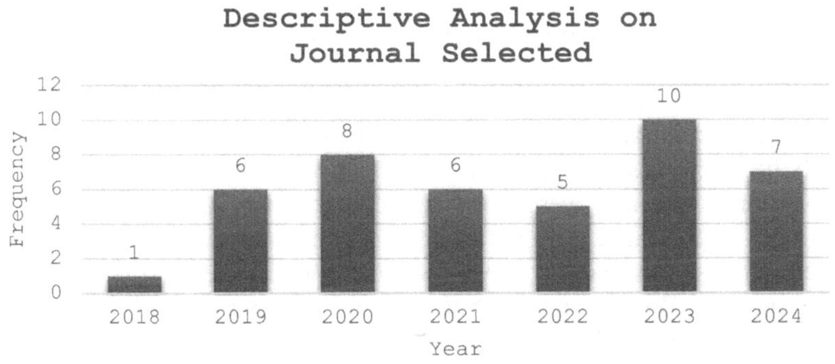

Fig. 7.3 Descriptive analysis on journal selected. *Source* Developed for this study

7.4.3 Descriptive Analysis

The publication years of the 43 selected publications are arranged chronologically in Fig. 7.3. Notably, scholarly writings about the gig economy significantly increased in 2019. This analysis only took into account publications published before August 2024. Furthermore, the 43 studies were published in several journals, meeting the quality standards for ranking journals. In the meantime, two articles were contributed by each of the succeeding journals: Annual Review of Organizational Psychology and Organizational Behavior; Human Relations; Information Technology & People; International Journal of Manpower; Journal of Asian and African Studies; Labor Economics and Safety science. One article each about gig workers' difficulties, motivations, or welfare was published in the remaining twenty-nine journals. The distribution of research methodologies was found to heavily rely on quantitative methods. Out of the forty-three papers, twenty (46.51%) paper used a quantitative technique, while eight (18.60%) accounted for paper used a qualitative approach, twelve (27.91%) papers chose a conceptual approach, and the remaining 3 (6.98%) papers used a mix-method.

7.4.4 Thematic Analysis

The study in this paper sought to identify themes from the data collected from 43 documents. The emotional challenge, identity challenge, organizational challenge, relational challenge, viability challenge, mental workload, emotional workload, physical efforts, job autonomy, competence, relatedness, work-life quality, job satisfaction, and life quality are recurrent concepts associated with gig work identified through initial coding. Every researcher produced an initial theme during the several rounds of review and improvement that we carried out (Naeem et al. 2023). After that, there was a collaborative discussion on these themes to address discrepancies and guarantee thorough data coverage. Based on these preliminary themes, we developed six aggregate dimensions that extensively comprehend gig employment. The categories encompass the issues encountered by gig workers, including economic, organizational, identity, relational, and emotional challenges. At the same time, the job demands for gig workers included emotional strain, mental workload, and emotional workload. Job relatedness, competency, and autonomy have all been discussed in job resources. The remaining themes were left as separate dimensions and were not integrated into any of the previously stated dimensions, as presented in Fig. 7.4. To make the analysis transparent and rigorous, we used to survey and record various draft and changes. Research conferences were organized to discuss and combine the results, which enhanced the credibility and accuracy of the study. The research led to general understanding of the gig economy by coming up with six aggregate dimensions. These points outline the areas to be further investigated and intervened and give valuable information about how complicated gig labor is.

Such classification is based on hierarchical relations between themes that are determined in the literature. The grouping of these themes was done in a repetitive manner and entailed mapping the influence of various challenges or job requirements on the important outcomes. These relationships are captured in the last conceptual model presented in Fig. 7.5 as it depicts the relationship between the six aggregate dimensions within a gig work ecosystem. It gives a basis upon which future theoretical and empirical investigation will take place.

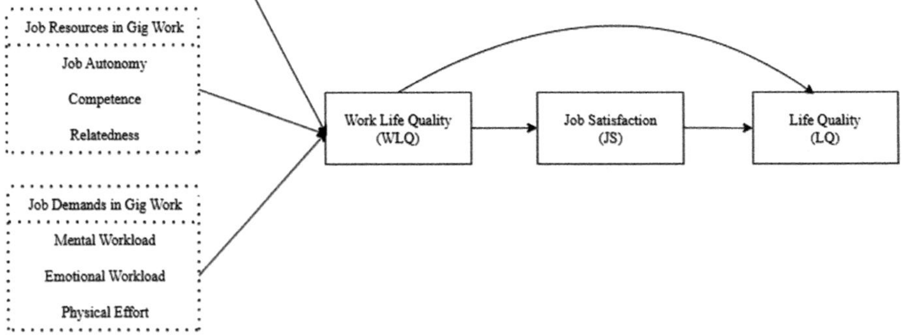

Fig. 7.4 Thematic analysis on journal selected. *Source* Developed for this study

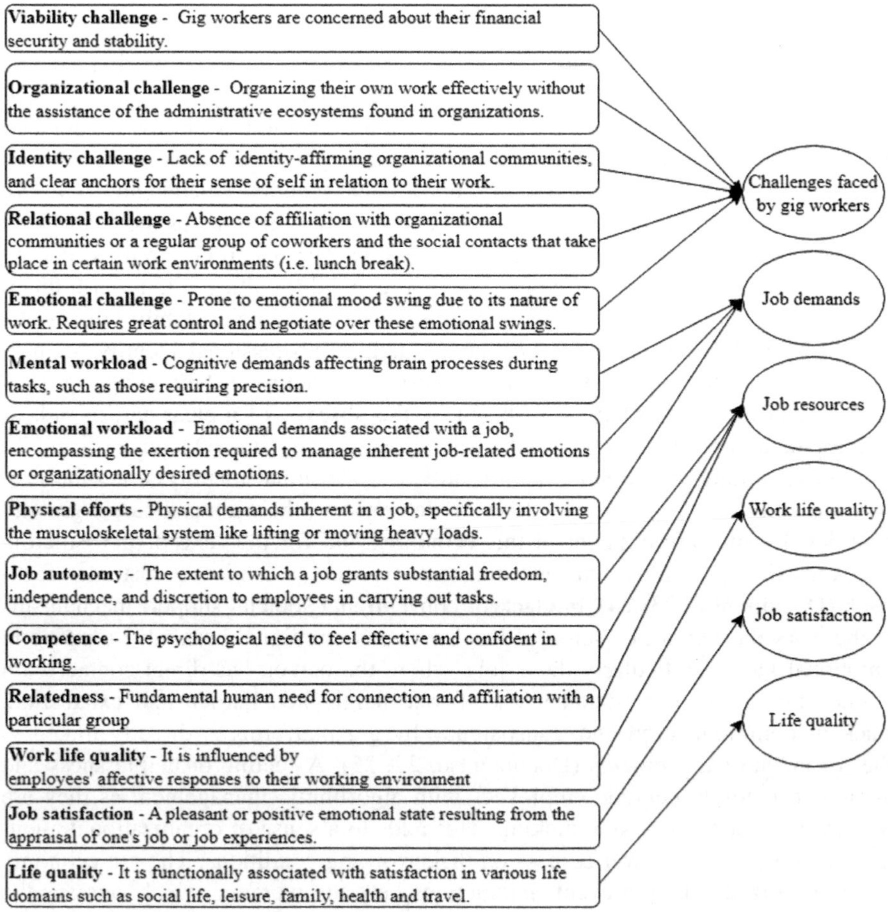

Fig. 7.5 Conceptual framework based on thematic analysis. *Source* Developed for this study

7.5 Findings and Discussions

Grounded by the tabulated findings beneath every category based on the thorough evaluation of all 43 papers, this section offers specific discoveries and deliberations on each category in the gig economy sector. The discussion is attributed to the conceptual framework of the given study as it includes the identified problems, work needs, work resources, work-life quality, work satisfaction, and general life quality.

7.5.1 Challenges of Gig Work

Gig economy presents various challenges to workers which can be divided into viability, identity, emotional, organizational, and relational challenges. The readings highlight compelling interest respecting to the economic viability of gig labor. Monetary uncertainty is a common problem among the gig workers because of inconsistent earnings and the lack of work-related benefits (Document 1, 15, 18). Gig workers at times have difficulties in having a stable income as the availability of jobs is not constant and the demand is unstable. The instability on the economy may be manifested in form of financial pressure and difficulties in long-term financial planning, which have implications on the overall financial stability of people. The fact that freelance jobs do not have a definite professional identity requires assistance to a large percentage of gig workers in defining their unique professional identity. It is further exacerbated by the absence of a clear career path or development opportunities within the frameworks of gig sites (Document 5 and 7). The move among jobs often leads to unified professional identity when the professionals obviously require assistance in determining their job and career growth. Emotional issues are eminent in this economy, and for the sake of the isolating attributes of gig employment and constant exposure to negative customer feedback (Document 8, 23, 24), the lack of conventional societal support mechanisms at the workplace creates a sense of loneliness and emotional burnout. Engaging in emotional labor, particularly those jobs where the person has direct contact with clients, increases the emotional load, pvoke heightened burden and exhaustion. Lack of communication and transparency by gig platforms is directly linked to the organizational problems (Document 6, 22, 25). A feeling of helplessness and alienation tends to pervade employees with algorithmic management as they are not involved in any decision making. This adds to a sense of organizational alienation and a sense of powerlessness on their working conditions. The gig economy is decentralized, which prevents interactions between workers, which worsens the connections and cooperation (Document 4, 12, 16). The unstructured work environment minimizes chances of peer support, mentoring, and networking, leading to being professionally isolated and being out of touch with the rest of the community. The issues of viability and the necessity to be more financially stable and receive benefits, as well as the necessity to have more structured career development opportunities within the frameworks of gig platforms, are accentuated by the concerns of identity. As a way of addressing emotional and relationship challenges, platforms need to create more favorable environments, which facilitate community and provide mental health resources. Existence of organizational issues means that there is need to have transparent and inclusive management practices. The practices will make the workers feel stronger and enhance their attitude of engagement and sense of belonging. A number of publications over the recent years have explored the particular challenges facing gig workers in their line of employment as given in Table 7.1.

Table 7.1 Summary of challenges from reviewed article. Source: Developed for this study

Document	Types of challenges	Discussion
1	Viability	Lots of gig workers complain about the inability to maintain their jobs because of the inability to predict demand and lack of work security
15		The platform-based work is not as stable as conventional jobs and majority of the workers cannot estimate how much they will earn in a week
18		Non equilibrium in earnings also often causes problems with emergencies or long-term savings by workers because of their discontinuous income
5	Identity	Workers articulate a sense of being marginalized in the job market, lacking a distinct professional trajectory or prospects for advancement
7		Gig work's nature causes them to struggle to precisely define their professional identity
8	Emotional	Gig workers can also be emotionally exhausted when they experience negative comments and reviews by customers regularly
23		Employees who are on gigs experience high rates of stress because of unforeseen workloads and lack of support
24		Gig work is immensely challenging to people who need to seek new employment relationships constantly, which is accompanied by emotional burnout and turbulent relationships
6	Relational	The increasing separation of gig employees and the companies that employ them causes the loss of loyalty, which may cause relational issues in the workplace
22		Gig work facilitates isolation which creates huge relationship problems since the workers have no chance to develop meaningful relationships at work
25		Lots of gig employees state that operators of platforms do not communicate and do not provide transparency about the job and earnings
4	Organizational	Absence of a systemized working situation in the gig work reduces chances to seek peer support and mentoring
12		The high physical demands of long working hours and lifting heavy objects have been reported by many gig workers, specifically within the delivery services
16		The competitive attributes of the gig economy may prevent the formation of cooperative relations among the workers

7.5.2 Job Demands of Gig Work

The job demands in the gig are divided into physical work, cognitive work, and emotional workload. Huge physical effort is meted out to the gig workers who are involved in the delivery and transportation services (Document 12, 13, 41). The physical tension caused by long working hours, dragging heavy items and extended sitting cause some health problems, in this case, musculoskeletal problems. The physical activity may have long-term health effects, reducing the level of job satisfaction and life quality in workers. Gig work often requires a great mental load due to the necessity of performing several tasks at once and time management as well as decision-making under great pressure (Document 20, 30, 38). The necessity to deal with traffic, multiple responsibilities, and respond to the needs of clients in a limited time results in cognitive exhaustion, which has a unfavorable influence on the mental state and productivity of employees. The emotional stress impact of gig jobs is the strongest in the jobs where there is direct contact with customers, which will be reflected in documents 3, 14 and 19. This emotional strain that might lead to burnout and job dissatisfaction can be achieved by addressing difficult clients, successfully creating and managing expectations, and managing negative feedback effectively. The demanding attributes of work within the gig economy are what makes it impossible not to provide more support networks that alleviate the physical, mental, and emotional challenges that workers encounter. The workload might be decreased through ergonomic adjustments and making frequent rest periods. To reduce mental burden, training and improvement of task management tools might be needed. The measures that might be used in order to address emotional stress are the extension of access to mental health resources and the creation of a more useful customer contact framework. Table 7.2 views the articles that have explored the job demands on the gig workers.

7.5.3 Job Resources of Gig Work

The job resources that are accessible to the gig workers are autonomy, relatedness, and competency. Gig labor is typically defined by a rather high degree of workplace freedom, where the workers are free to make their own schedules and places to work (Document 17, 26, 34). This flexibility is seen by gig workers as a great benefit since they can balance work and family life. Despite the inherent freedom of gig work, there are a number of platforms able to create a sense of community and connectedness through the use of internet forums and support groups (Documents 9, 10, 27). Workers who establish a strong relationship with the other workers usually show increased work satisfaction and commitment. The ability to make opportunities of skills development and education is important in enhancing the employee perception of proficiency (Document 21, 35, 42). Skills and happiness of gig workers may be enhanced by platforms that offer training and development tools, which will result in increased engagement and job fulfillment. The job resources are significant in the improvement of pleasure and engagement

Table 7.2 Summary of job demands from reviewed article. *Source* Developed for this study

Document	Types of job demands	Discussion
12	Physical effort	Most gig workers, especially in the delivery services complain of a lot of physical workload because of long hours and carrying
13		Gig workers, particularly the physically demanding ones, often complain of physical strain caused by excessive working hours and lack of break
41		Gig work has physical requirements, particularly in the delivery and manual labor, which heavily affect the workers
20	Mental workload	Multitasking, time management, and pressure to make decisions are some of the mental workloads that gig workers have to contend with
30		Gig workers usually have a heavy mental burden to negotiate between odd work schedules, multifaceted task demands, and to constantly find new jobs
38		The requirement to go through traffic and manage two or more tasks at the same time imposes a great cognitive burden on gig workers
3	Emotional workload	Job demands such as emotional workload and mental stress are a significant burden for gig workers, leading to high rates of burnout
14		Emotional labor is a significant part of gig work, especially in roles that involve direct customer interaction
19		The psychological impact of having a job and not having a stable job with all the perks is severe to most of the gig workers, which results in burnouts and job dissatisfaction

among the workers in gig. High degree of job autonomy in gig labor is one of the major strengths and is a plus because it offers flexibility and control of work hours. A relationship should be established and opportunities to improve skills must be offered to make sure that gig workers do not feel unsupported and incompetent in their positions. Platforms should focus on the enhancement of these resources to enhance employee satisfaction and strengthen the retention of staff. The research studies summarized on employment resources experienced by gig workers were compact in Table 7.3.

7.5.4 Work-Life Quality (WLQ)

Gig workers also struggle with the inability to enjoy a good balances between work and activities outside of work because of unpredictable working schedules and irregular revenue (Document 11, 28, 39, 40). The flexibility of gig work may be a benefit or a detriment in a situation, depending on the case. Though a number of the employees enjoy the freedom of choosing their schedules, there are those who have problems in maintaining a distinction between work life and individual life. The inherent instability of the gig job is quite a challenge to the employee

Table 7.3 Summary of job resources from reviewed article. *Source* Developed for this study

Document	Types of job resources	Discussion
17	Job autonomy	Traits of gig work that are highly appreciated are autonomy because gig workers hold the flexibility to choose task at the times and tasks of their choice
26		Gig workers enjoy one of the most significant benefits in the form of high autonomy, as it facilitate them to regulate their working roster and conditions
34		The element of autonomy is a motivational aspect, which implies the ability to perform a particular activity, whereas competence refers to the effectiveness of implementing a specific task
9	Relatedness	The issue of relatedness is still present in the background of gig work because numerous of them work alone with minimal communication with their peers and employers
10		Feelings of relatedness among gig workers can be promoted by platforms that motivate worker interaction and collaboration
27		It has been recognized as an essential resource to maintain engagement and reduce isolation that many workers experience due to the development of social connections and relatedness in the form of gig work communities
21	Competence	Competence development in gig work is largely self-directed, as workers are expected to continuously upskill in the interest of maintain competing in the market
35		Competence is a key driver in motivating workers to upskill themselves, contributing to the gig economy's resilience in changing labor markets
42		Competence and self-leadership are essential resources for gig workers, as they must manage their own professional development without support from a traditional employer

in terms of achieving a reliable income, which hinders their ability to balance personal and professional life properly. In order to improve the WLQ, there is a necessity to address the problem of variable earnings and introduce more organized employment changes. The platforms must make the work tasks more predictable and must look at ways of providing financial support during the low demand seasons. The composition of this study on the WLQ of gig employees is presented in Table 7.4.

7.5.5 Job Satisfaction (JS)

The level of job satisfaction is altered over different consideration, including job autonomy, remuneration, and meaningful work, among others, among gig workers. The perception of independence and the ability to control the timetables by workers is significantly related to JS (Document 33, 35). Those employees who think

Table 7.4 Summary of WLQ from reviewed article. *Source* Developed for this study

Document	Discussion
11	Although gig work grant flexibleness, the lack of financial stability and long-term prospects often negatively impacts WLQ
28	The gig economy's flexible nature allows individuals to balance their responsibilities, though this depends on the type of platform and work arrangements provided
39	Work-life balance refers to time, involvement, and satisfaction ratio, showing individuals who invest more time in family experience a higher life quality
40	The characteristics of gig employment which is difficult to predict have hindered those workers from achieving a satisfactory work-life equilibrium

Table 7.5 Summary of JS from reviewed article. *Source* Developed for this study

Document	Discussion
31	Gig employees susceptible to MTurk as a main stream of salary reported lower JS compared to those who did not depend on it for income.
33	Autonomy and flexibility are closely associated with gig work, which can be compromised in case of unpredictability of work and short of compensation package and work protection.
35	Gig workers who enjoy autonomy and have autonomy over their schedules report higher levels of JS.

that they are in control of their job contracts can have better JS. Paying proper amounts of money is a crucial element that determines work satisfaction within gig economy (Document 31). To improve on the condition of JS among gig workers, the platforms ought to focus on providing reasonable compensation, significant employment opportunities, and maintaining high levels of working autonomy. The results of the research on gig workers JS are as indicated in Table 7.5.

7.5.6 Life Quality (LQ)

LQ of gig workers tends to deteriorate because of financial instability, lack of social protection, and work pressures. Document 29 has highlighted the impact of financial instability on the LQ to a large extent. The inherent uncertainty of contemporary nature of work and the absence of conventional job benefits intensify stress and anxiety and therefore reduce the total LQ. However, some people experience the decline in their LQ of the lack of structured working conditions and the uncertainty of their income. The LQ of gig employees is adversely influenced by the insufficient of protection of adequate protection in society (Document 32). Individuals lacking access to these precautions often encounter increased stress and reduced LQ. Enhancing the LQ for gig workers requires addressing their financial insecurity and guaranteeing access to social safety nets. Platforms and authorities

Table 7.6 Summary of LQ from reviewed article. *Source* Developed for this study

Document	Discussion
29	Gig workers often report a mixed impact on LQ, where increased flexibility improves it, but income volatility and job insecurity negatively affect long-term LQ
32	Gig workers report fluctuating levels of LQ, as the ability to carry out task on their own terms' contrasts with the financial instability and the limited access to healthcare and other social protections
36	Employee satisfaction frequently indicates less absenteeism, more creativity, and better LQ. Poor organizational culture reduces motivation, work quality, and overall LQ
37	LQ is impacted by the evolving nature of work, where job insecurity and work intensity often conflict with LQ and long-term career sustainability

can consider implementing measures that offer gig workers increased stability and support (Document 36, 37). Improving these aspects will lead to a superior LQ for gig workers. Table 7.6 has shown results on the LQ of gig workers.

7.6 Conclusion

This conceptual paper devotes to the state of knowledge on gig employment due to its intensive and multidimensional experiences of low-skill physical workers of gigs who are regularly underrepresented in the gig economy literature. The study further establishes important challenges (viability, identity, relational, emotional, and organizational) to be experienced by gig workers in terms of job demands (emotional workload, mental workload, physical effort), and job resources (job autonomy, competence, relatedness) that all have an overall impact on the work-life quality (WLQ), job satisfaction (JS), and life quality (LQ). The main theoretical contribution of the present paper consists in its utilization of the job demands-resources (JD-R) model into the background of gig work. Although the JD-R model were used in the past in those structured workplace environments, in this case, it was modified to fit the non-standard and platform-mediated employment environment, demonstrating how the gig-specific employment characteristics enhance work annoyance and burnout risk, including personal income instability, absence of organizational support, and algorithmic control. In contrast, such resources as flexibility of schedules and perceived autonomy provide partial protection against these pressures. We realize however that our findings suggest the resources tend to not be adequate to offset the uncertain environment of gig work in low skill physical jobs.

Organizing the results into a conceptual framework, upon the foundation of JD-R model, the paper has presented a theoretical map that describes the processes through which job demands and resources combine influence toward the perceived level of life quality. It points to the importance of balance, when the improvement of job resources will not be enough to improve precarity-related challenges when they have not been addressed to the roots. With this strategy, this will motivate emerging scholars to redefine the application of the JD-R model in the non-traditional employment context and customize it to the fluid, unregulated, and platform-based working conditions. In practical terms, the research has interesting information, which is important to platform providers, labor unions, and policymakers. In the case of platform companies, there also is the opportunity to comprehend to what degree different job demands and resources can affect the workers' overall well-being. The sphere of action is advised to be taken by the labor unions who can introduce gig worker input into the policies used by the platforms and foresee missteps in considering design patterns. Policymakers are on the other hand advised to come up with protective measures addressing the origin of instability and make platform management transparent and accountable.

Overall, the paper has a substantial theoretical contribution in adapting the JD-R model to the actual conditions of the economy of gig. It goes beyond the measure of productivity and focuses on the contented state of gig workers and can form a starting point and provide a theoretical level of future research and policy formulating. Although this study is conceptual in nature, it provides very important fundamental research in terms of empirical validation, particularly in a wide range of nations. The future research should investigate the ways particular job demands and resources could emerge in various gig platforms and how they could be addressed to help improve workers WLQ, JS, and overall LQ in the changing working environment of platform work.

APPENDIX A

No	Author	Document Title
1.	Kassim et al.	Embedding gig economy in Malaysia higher education: The case of Universiti Teknologi MARA
2.	Abdul Rahim et al.	Strengthening the gig economy: Future of digital labor workforce platform post-COVID-19
3	Daud et al.	Adapting to the gig economy: Determinants of financial resilience among "Giggers"
4	Radzi et al.	Protecting Gig Workers' Interests in Malaysia Through Registered Association Under Societies Act 1966
5	Amin	Balancing the right of gig economy workers in the context of collective bargaining
6	Ashford et al.	From surviving to thriving in the gig economy: A research agenda for individuals in the new world of work

(continued)

(continued)

No	Author	Document Title
7	Wu et al.	Labor control in the gig economy: Evidence from Uber in China
8	Woon et al.	Impact of JD-R model on psychological Well-being of gig workers
9	Gandhi & Sucahyo	Architecting an advanced maturity model for business processes in the gig economy: A platform-based project standardization
10	Chen	Are we only all by ourselves? A double-level perspective to cope with the insecurity of the nonstandard gig work model
11	Hunt & Samman	Domestic work and the gig economy in South Africa: Old wine in new bottles?
12	Christie & Ward	Delivering hot food on motorcycles: A mixed method study of the impact of business model on rider behaviour and safety
13	Umair et al.	Understanding the influence of technostress on workers' job satisfaction in gig-economy: An exploratory investigation
14	Tan & Low	Do gig workers enjoy work-life balance? Exploring the influencing factors for work-life balance among the gig workers
15	Schwellnus et al. (2019)	Gig economy platforms: Boon or bane?
16	Ahmad et al.	Gig economy: The promise of platform work
17	Jabagi et al.	Gig-workers' motivation: Thinking beyond carrots and sticks
18	Uchiyama et al.	Gig workers, social protection and labor market inequality: Lessons from Malaysia
19	Zaman et al.	Having a whale of a time: Linking self-determination theory (SDT), job characteristics model (JCM) and motivation to the joy of gig work
20	Mohsin et al.	Growing gigs: A conceptual report on job autonomy and work engagement on gig workers' performance
21	Ahmad	Gig economy: The future of working
22	Kurian et al.	Navigating the gig economy: Exploring challenges and motivations for the wellbeing of gen Z gig workers
23	Kwan & Low	Joy of gig work?! A study of Malaysian gig workers
24	Kaine & Josserand	The organisation and experience of work in the gig economy
25	Ahmad	Mapping gig workers as the new economy post COVID-19
26	Kuhn et al.	Human resource management and the gig economy: Challenges and opportunities at the intersection between organizational HR decision-makers and digital labor platforms
27	Carroll	Factors that affect perceptions of gig workers
28	Taylor et al.	Physical and psychological hazards in the gig economy system: A systematic review
29	Bracha & Burke	How big is the gig? The extensive margin, the intensive margin, and the hidden margin
30	Norlander et al.	The effects of technological supervision on gig workers: Organizational control and motivation of Uber, taxi, and limousine drivers

(continued)

(continued)

No	Author	Document Title
31	Keith et al.	Mechanical Turk and the gig economy: Exploring differences between gig workers
32	Cropanzano et al.	The organizational psychology of gig work: An integrative conceptual review
33	Apouey et al.	Gig workers during the COVID-19 crisis in France: Financial Precarity and mental Well-being
34	Mahmod & Hassan	Spatial crowdsourcing: Opportunities and challenges in motivating Malaysian Women's participation in gig economy
35	Tara & Iqbal	Examining the impact of job demands, resources and technostress on psychological wellbeing of gig workers: A theoretical model
36	Norwani et al.	The influence of gender, age, academic level and motivation on satisfaction toward benefits received by gig economy workers
37	Fraccaroli et al.	Challenges in the new economy: A new era for work design
38	Binu Raj et al.	Faculty engagement, quality of work-life, organizational commitment and spiritual leadership-examining the mediation and moderation effects
39	Picatoste et al.	Job quality and Well-being in OCED countries
40	Warhurst & Knox	Manifesto for a new quality of working life
41	Daniels	Navigating Precarity: Health and safety challenges in Southeast Asia's gig economy food delivery sector
42	Hassan et al.	The role of self-resilience on the relationship between self leadership and employee engagement: A conceptual paper
43	Li et al. (2024)	Work engagement of online car-hailing drivers: The effects of platforms' algorithmic management

References

Ahsan M (2020) Entrepreneurship and ethics in the sharing economy: a critical perspective. J Bus Ethics 161(1):19–33. https://doi.org/10.1007/s10551-018-3975-2

Auguste D, Roll S, Despard M (2022) Democratizing the economy or introducing economic risk? Gig work during the COVID-19 pandemic. Work Occup 50(1):3–32. https://doi.org/10.1177/07308884231202032

Business Research Insights (2024) Gig economy market size, share, growth, and industry analysis, by type (asset-sharing services, transportation-based services, professional services, household & miscellaneous services [HGHM], others), by application (traffic, electronic, accommodation, food and beverage, tourism, education, others), regional forecast by 2031. Business Research Insights https://www.businessresearchinsights.com/market-reports/gig-economy-market-102503

Cloonan M, Williamson J (2023) Musicians as workers and the gig economy. Pop Music Soc 46(4):354–370. https://doi.org/10.1080/03007766.2023.2206485

Dazzi D (2019) Gig economy in Europe. Ital Labour Law e-J 12(2):67–122. https://doi.org/10.6092/issn.1561-8048/10785

Erlich M (2021) Misclassification in construction: the original gig economy. ILR Rev 74(5):1202–1230. https://doi.org/10.1177/0019793920972398

Ghazali AJB, Muhammad NMBN (2021) An overview of gigs economy in Malaysia. Nasrul Aiman Bin Abd Aziz Web Des 162

Harith NHM, Kamarunzaman NZ, Ismail TNT, Zainuddin A, Binti Jamil NFA (2022) A preliminary findings of urban poor Malay households livelihood strategies in Klang Valley, Malaysia. Res Militaris 12(2):2254–2271

Hasija S, Padmanabhan VP, Rampal P (2020) Will the pandemic push knowledge work into the gig economy? Harv Bus Rev. https://hbr.org/2020/06/will-the-pandemic-push-knowledge-work-into-the-gig-economy

Huđek I, Širec K (2023) The terminology and the concept of the gig economy. Ekonomski Pregled 74(1):34–58

Kidd SC (2020) Conceptualizing the gig economy in Malaysia. Res Soc Adv (REFSA) 7

Koutsimpogiorgos N, Van Slageren J, Herrmann AM, Frenken K (2020) Conceptualizing the gig economy and its regulatory problems. Policy Internet 12(4):525–545. https://doi.org/10.1002/poi3.237

Lepanjuuri K, Wishart R, Cornick P (2018) The characteristics of those in the gig economy. Department for Business, Energy and Industrial Strategy. https://www.gov.uk/government/publications/gig-economy-research

Levitt L (2021) Sex work/gig work: a feminist analysis of precarious domina labor in the gig economy. In: The gig economy. Routledge, pp 58–71. https://doi.org/10.4324/9781003165813

Loganathan M (2022) Assessing social sustainability in the gig economy. Indian J Labour Econ 65(3):831–843. https://doi.org/10.1007/s41027-022-00411-5

Makhtar M, Ghadas ZAA, Haque M (2022) Analysis of workers' rights in the on-demand economy with special reference to vertical inequality. IIUMLJ 30:129–152

Mohamed Shaffril HA, Samsuddin SF, Abu Samah A (2021) The ABC of systematic literature review: the basic methodological guidance for beginners. Qual Quant 55:1319–1346. https://doi.org/10.1007/s11135-020-01059-6

Naeem M, Ozuem W, Howell K, Ranfagni S (2023) A step-by-step process of thematic analysis to develop a conceptual model in qualitative research. Int J Qual Methods 22:16094069231205789. https://doi.org/10.1177/16094069231205789

Oei SY (2018) The trouble with gig talk. Law Contemp Probl 81(3):107–136. https://scholarship.law.duke.edu/lcp/vol81/iss3/5

Page MJ, Moher D, Bossuyt PM, Boutron I, Hoffmann TC, Mulrow CD, Shamseer L, Tetzlaff JM, Akl EA, Brennan SE, McKenzie JE (2021) PRISMA 2020 explanation and elaboration: updated guidance and exemplars for reporting systematic reviews. BMJ 372

Parigi P, Ma X (2016) The gig economy. XRDS: crossroads. ACM Mag Stud 23(2):38–41. https://doi.org/10.1145/3013496

Rethlefsen ML, Page MJ (2022) PRISMA 2020 and PRISMA-S: common questions on tracking records and the flow diagram. J Med Libr Assoc 110(2):253–257. https://doi.org/10.5195/jmla.2022.1516

Schwellnus, C., Geva, A., Pak, M., & Veiel, R. (2019). Gig economy platforms: boon or bane? OECD Economics Department Working Papers, No. 1550. https://doi.org/10.1787/fdb0570b-en

Slee T (2017) What's yours is mine: against the sharing economy. OR Books

United Nations (2020) SDGs still offer best option to reduce worst impacts of COVID-19 and to recover better Department of Economic and Social Affairs. https://www.un.org/en/desa/sdgs-stilloffer-best-option-reduce-worst-impacts-covid19-and-recover-better

Vallas S, Schor JB (2020) What do platforms do? Understanding the gig economy. Annu Rev Sociol 46:273–294. https://doi.org/10.1146/annurev-soc-121919-054857

Woodcock J, Graham M (2019) The gig economy: a critical introduction. Polity Press

Zutshi A, Grilo A (2019) The emergence of digital platforms: a conceptual platform architecture and impact on industrial engineering. Comput Ind Eng 136:546–555. https://doi.org/10.1016/j.cie.2019.07.025

Siao Ying Choo received her Bachelor of International Business (Hons) with first class, and she is currently a MPhil student in UTAR. Her current research interests and research topic is focus on the gig economy which is the new way of working nowadays. Her main supervisor is Dr Low who is also the second author for this paper while co-supervisor is Dr Har who is the third author for this paper.

Mei Peng Low is currently an Associate Professor of Faculty of Accountancy and Administration in UTAR. Her research interests focus on Organisational Behaviour, Micro-level CSR, Ethics and Social Responsibility and Human Resource Management. She complements her teaching experiences with actively involve in research work, journal publication and consultancy projects. She has published in leading journals such Business Ethics, the Environment and Responsibility, Quality and Quantity, Social Responsibility Journal and others.

Wai Mun Har received his Bachelor of Economics (Hons) and Master of Economics from Universiti Kebangsaan Malaysia (UKM). He received his Doctor of Philosophy (PhD) from Universiti Putra Malaysia (UPM). Currently, he is a senior lecturer at the Economics Department, Faculty of Accountancy and Management, UTAR and has teaching experiences of more than 19 years. His publications include books, book chapters, academic journals, periodicals, and newspapers.

Part IV

Sustainability, Empathy and Well-Being

8 Exploring the Linkage Between Decent Work-Derived Mental Health Wellbeing and Workforce Performance in Brickfield Sector

Manidipa DasGupta and Sumanta Majumder

Abstract

A healthy working atmosphere is the fundamental condition of a decent work environment (SDG-8) which, confirming workers' health-related wellbeing (SDG-3), is supposed to be the right. Health-related wellbeing, connected especially with the psychological (mental) aspect, is likely to be linked with workers' productivity at the workplace. The present empirical study hence aims at exploring the linkage between decent work-derived mental health wellbeing and workforce performance. To fulfil the objective, it collected responses on 281 randomly selected brickfield workers of West Bengal on NIOSH-WellBO scale. With structural equation modelling (SEM) through SPSS-AMOS (version 23), it finds that 1% change in decent work-derived mental health wellbeing generates 31% change in work performance. The study proves its policy significance for reflecting the voice of the workforce of a risk-prone enterprise for decent standardisation of the working environment in view of enhancing the workers' mental health wellbeing and work performance.

Keywords

Decent work • SDG 8 • Mental health • Wellbeing • Workforce performance

M. DasGupta
The University of Burdwan, Bardhaman, West Bengal, India
e-mail: mdgupta@com.buruniv.ac.in

S. Majumder (✉)
Bhatter College, Dantan, West Bengal, India
e-mail: sumantamajumder27@gmail.com

B. Kaur et al. (eds.), *Integrating Industry 5.0 and 4.0 for Technology and Humanity*, Contributions to Environmental Sciences & Innovative Business Technology,
https://doi.org/10.1007/978-981-92-1780-9_8

8.1 Introduction

Work and health have a two-way relationship that concerns occupational health wellbeing (ILO 2015; Chari et al. 2022; Su and Chan 2023) and confers decent work conditions at workplace, i.e. (i) fair income; (ii) safety-security in employment and working condition; (iii) equal opportunity and treatment for all; (iv) social protection for the workers and their families; (v) personal development and social integration; and (vi) workers' freedom to express (ILO 1999). Disappointedly, the concerned rhythm is found disorganised for imbalanced work-life status and unhealthy working environment which cause workers' occupational illbeing. ILO in its recent published report (2023) released that 2.3 million people around the world every year used to surrender to work-related accidents or diseases linking with over 6000 deaths every single day, while mental illness due to depression was found to hold 8% of the global burden (WHO-ILO: WASH 2018–25 Agenda).

Referring to the world scenario, the developing countries are traced to bear more than 80% of the global burden of occupational disease and injury where India is found at high risk (WHO-World Bank 2007; Das 2020). It is detected severe mainly in unorganised sector where enterprises due to infrastructural bottleneck, financial lacuna, absence of standardised health-safety policy and unawareness of workers are ever traced distracted from the decent work condition at workplace and thus suffer from illbeing in workforce for both physical and mental health issues (Alam et al. 2021; Rajan 2022). Amongst the unorganised sectors, the workers of Brickfield manufacturing are found experiencing acute occupational hazards and work-related injuries mainly due to its inherent hazardous working environment, extreme workload, tough working schedule, heat stress, dust exposure, ergonomic issues, etc. (Das 2015; Kumari 2018; Bajracharya et al. 2021; Bansal et al. 2022). Resultantly, the workers used to suffer from morbidity for headache, body aches, problems in vision, cough, breathlessness, hearing loss, stomach and throat cancer, stress, sadness, etc. (Kumari 2018; Saha et al. 2020; Nanjunda and Venugopal 2022). The sector is not regulated by any Government legislation for which no standardised movement is found to protect the workers from occupational hazards (National Brick Mission 2016).

But, despite having tremendous health hazard and lack of standardised preventive measures, the brick workers genuinely promote a hike in sectoral production. It was substantiated by the recent World Bank Report (2020) where India was found as the second largest manufacturer of bricks after China with 250 billion annual production and 15 million employment generation. The Gangetic plains (Assam, Bihar, Haryana, Punjab, Uttar Pradesh and West Bengal) account for India's 65% brick production and the rest 35% is concentrated in the peninsular and coastal India (World Bank 2020).

The promising potential of the enterprise in the economy along with the ongoing threat towards its subsistence performance due to lack of decent work environmental surface for employees there, motivates the present study to extend its findings on the research question like—*what is the relationship between decent work-derived mental health wellbeing and workers' performance in the Brick-field sector?*

Research on decent work has been developed after the ILO applied decent work for the first time in 1999. Subsequently, various studies found that there is a relationship between occupation and health (ILO 2015; Chari et al. 2022; Su and Chan, 2023). From the literature, it is also detected that several studies (McCaughey et al. 2013; Uddin 2021; Greiner et al. 2022; Bentley et al. 2023) extended their efforts in the domain of mental health-related wellbeing. Numerous (Hessels et al. 2017; Jones et al. 2019; Martins-Silva et al. 2020) studies considered the relationship between decent work and mental health wellbeing. But, no study was found to focus on the relationship between decent work environment-derived mental health wellbeing and workers' performance at the workplace, till the review. Moreover, no study focuses on the 'Brickfield' sector, which is considered significant in this context.

Hence, to justify the research question and fill the knowledge gap, the study primarily aims at detecting a relationship between workplace decent work-derived mental health wellbeing and workforce performance in Brickfield. More specifically, it wants to detect the relationship between the decent work components ('workplace policies and culture with workers', 'workplace safety measures with workers' mental health well-being' and 'workplace health and hygiene practices and workers') and mental health wellbeing of brickfield workers.

The study in this regard has made its effort to extend an in-depth micro-empirical analysis in the western part of West Bengal (sixth largest in India) covering four districts (Bankura, Purulia, West Midnapore and Hooghly) of it with maximum proportion of brick manufacturing of the state there. Keeping in view of the background of the study, four more sections are framed. In Sect. 2, it highlights the 'Theoretical Framework and Hypothesis Formation'. In Sect. 3, the 'Methodology' is addressed, while in Sect. 4, the 'Results and Discussion' parts are highlighted. Section 5 discloses the 'Conclusion and Recommendations'.

8.2 Theoretical Framework and Hypothesis Formation

Workplace physical environment and safety climate, synonymously known as Occupational Safety and Health (OSH), is referred to the discipline which deals with the prevention of work-related injuries and diseases and also to the protection and promotion of the health of workers (Chari et al. 2022). WHO (2010) defined a healthy workplace as one in which workers and managers collaborate to use a continual improvement process to protect and promote the health, safety, wellbeing and sustainability of all workers. Thus, the poor working conditions lead to an indecent work environment. ILO (2015) in this regard promoted 'Safety + Health for All' to improve the health and safety of workers universally in all settings. Everyone has a right to get a safe and healthy life at work and thus a safe and healthy workforce is the foundation of sustainable development. Afterwards, the Vision Zero Fund (VZF) is integrated with this program to achieve zero fatal cases,

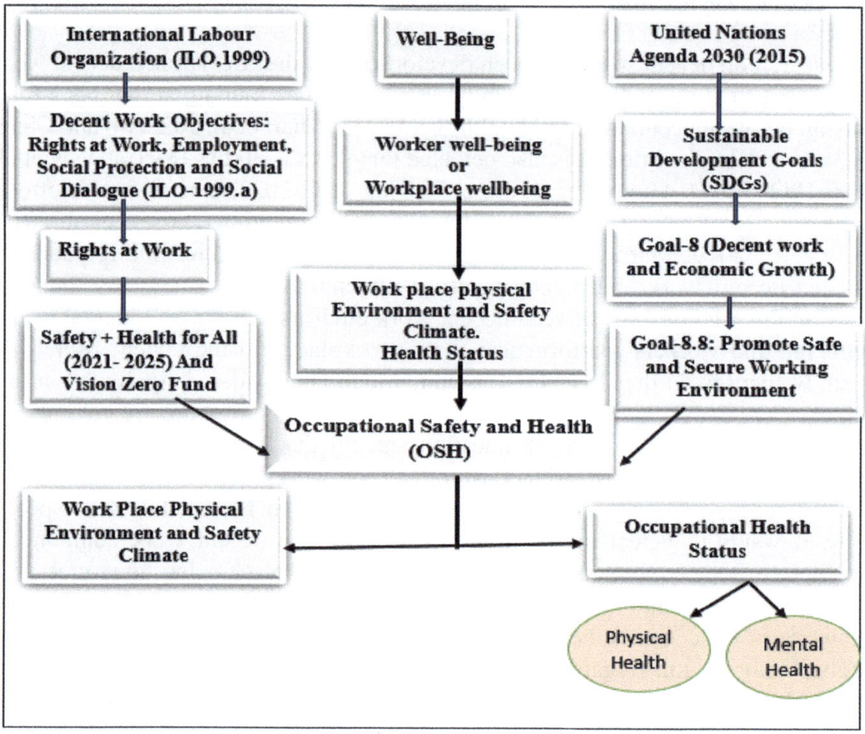

Fig. 8.1 Workplace physical environment, safety climate and occupational health status. *Source* Authors' own elaboration

injuries and diseases at the workplace. 'Safety + Health for All' (ILO Flagship Programme 2016–2020 and 2021–2025) contributed to the United Nations 2030 Agenda for Sustainable Development Goals directly through Goal 8 (decent work) [target 8.8 (safe working environment)] and indirectly through Goal 3 (healthy lives) [target 3.9 (hazardous environment)] and Goal 12 (responsible production and consumption) (Fig. 8.1).

8.2.1 Work Place Decent Work Environment and Workers' Mental Health

The perception of decent work environment is subjective because it depends on how each person visualises their own working conditions. Various studies (Hafee et al. 2019; Bayram 2019; Alam et al. 2022a, b) have found that decent work environment includes (i) Workplace Policy and Culture i.e., workplace culture, workplace respect, job recognition, employer trust, superior support, work values, earning satisfaction, (ii) safety measures, i.e. safety training, work safety resources, safety audit, safety equipment, safety climate, (iii) health and hygiene practices, i.e. safe drinking water, first aid, sanitisation viz. clean toilet and bathroom, clean basin and toilet soap and rest room. In this context, WHO (2010) provides four major areas which affect a healthy workplace, such as (i) the physical work environment, (ii) the psychosocial work environment, (iii) personal health resources and (iv) enterprise involvement in the community. Therefore, a set of positive decent working conditions influences workers' holistic wellbeing, including physical and mental conditions (Su and Chan 2023), living conditions, health-safety issues, social justice, equality, etc. (Webster et al. 2016; Dery and Bisung, 2022).

Considering the abovementioned fact, the following hypothesis can be framed.

$\mathbf{H_{01}}$: There is no significant relationship between the decent work components ('Workplace policies and culture', 'Workplace safety measures' and 'Workplace health and hygiene practices') and workers' mental health.

8.2.2 Mental Health Wellbeing and Workforce Performance

It is endorsed that healthy workers support organisational sustainability by reducing the chance of presenteeism (Horváthová et al. 2023). In this context, mental health wellbeing at workplace holds special significance (Obrenovic et al. 2020) as work-related mental illbeing like work stress, has negative impact on workers' performance, attention and cognition (Pandey 2020; Sari et al. 2021; Manjunatha 2023; Storbeck et al. 2024). Further, literature (Auten and Fritz 2019; Hennekam et al. 2021; Lu et al. 2022; Erwanto et al. 2024) argue that distressed mental health wellbeing is responsible for lower workers' participation in the enterprise where the unorganised sector (Manimekalai 2021) is found to suffer more than that of the organised one (Dery et al. 2023).

Wellbeing of workers finds direct relationship with inter-personal relation, interaction and collaboration amongst them (Mastroianni and Storberg-Walker 2014; Hafeez et al. 2019) where leaders' interference at work, ineffective policies, etc. (Monteiro and Joseph 2023) are negatively associated with it (Willett et al. 2023). Circumstantially, income potential of an enterprise is related to health care and high life satisfaction of employees (Bayram 2019; Mujtaba and Kaifi 2023). Following this wavelength, training and occupational safety measures have positive effects on the improvement of working conditions of employees and thus they carry direct relationship with improvement of workforce health status (physical and psychological) (Sari 2009; Oakman et al. 2018). The previous studies (Sari 2009; Arbogast et al. 2016; Abrampah et al. 2017; Alam et al. 2022a, b; On'gonge and Ngeno 2022; Johnson and Fujishiro 2023) also endorsed the effectiveness of sanitisation infrastructures, clean toilet system, hygiene practice and first aid availability at workplace on the quality of employees' mental health and productivity.

Considering the abovementioned fact, the following hypothesis can be framed.

H02: There is no significant relationship between decent work-derived mental health wellbeing and workers' work performance at the workplace (Fig. 8.2).

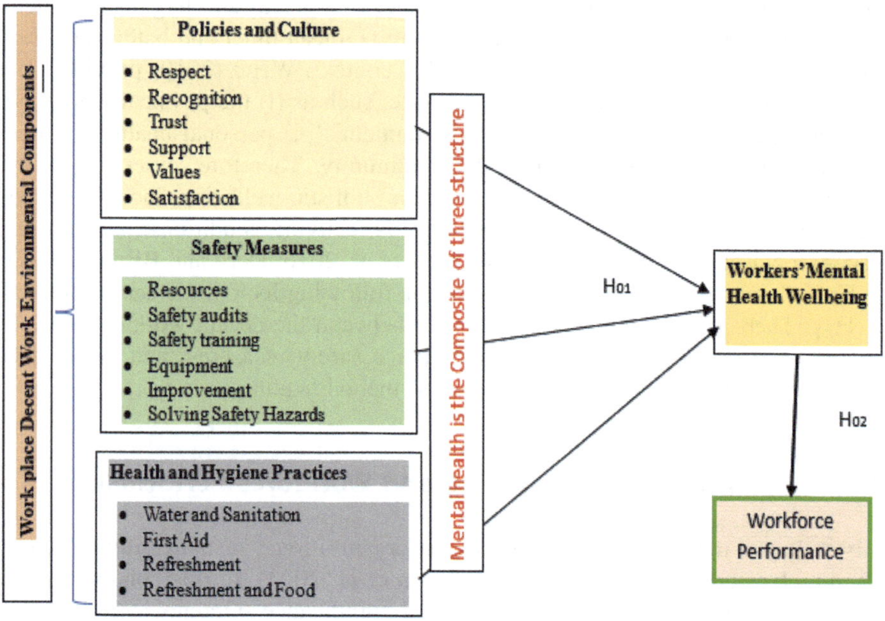

Fig. 8.2 Decent work-derived mental health wellbeing and workforce performance. *Source* Author's elaboration

8.3 Methodology

8.3.1 Data Collection

The present cross-sectional empirical study uses both primary and secondary sources of data. For primary data collection, it opts for four districts of western part of West Bengal (Bankura, Purulia, West Midnapore and Hooghly) from September 2023 to October 2023. Before the primary survey, a pilot survey was conducted on 50 respondents of Bankura district in August, 2023 to finalise the questionnaire. For secondary data, it adopts information from Government Reports (Ministry of Law and Justice), Reports from international organisations (ILO, WHO, etc.), unpublished dissertations, articles, newspapers, etc.

8.3.2 Sampling and Sample

The present study has purposely selected four districts of western part of West Bengal—(Bankura, Purulia, West Midnapore and Hooghly) covering 70% of brick kilns of the state at the bank of the Darakeswar River (Ghosh 2024). The fly-ash brick-making project was grounded at Bankura block 1 in 2015 which was an

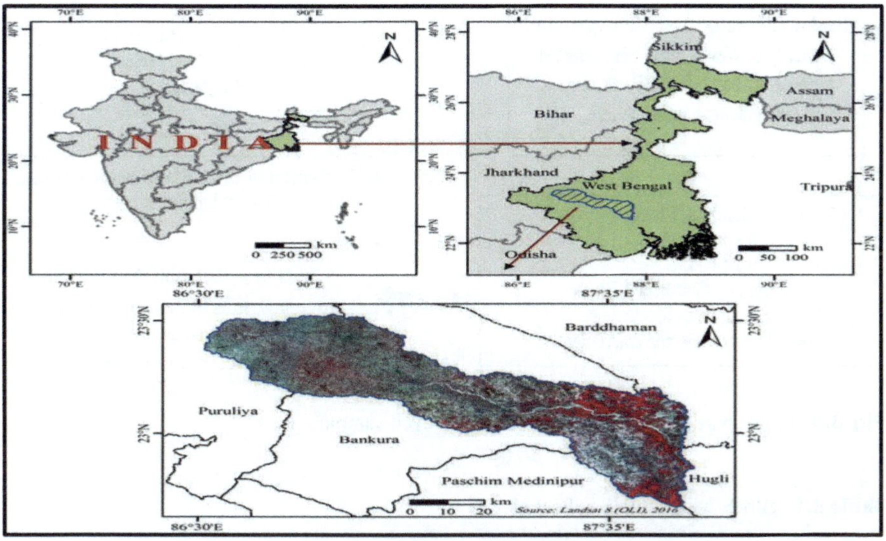

Fig. 8.3 Western part of West Bengal and bank of Darakeswar river area. *Source* www.researchg ate.net

ideal and innovative project set up under Mahatma Gandhi National Rural Employment Guarantee Act (MGNREGA) for the first time in West Bengal (Chatterjee 2016). From the four districts, five blocks viz. Kotulpur and Joypour (Bankura); Hura (Purulia); Garhbeta-1 (West Midnapore) and Arambagh (Hooghly) have been selected conveniently. From the selected blocks then 281 brickfield workers from 30 units have been selected randomly (9 units from Kotulpur, 6 units from Joypur, 4 units from each Hura and Garhbeta-1 and 7 units from Arambagh)[1] out of 1083 workers performing in a total of 49 units (DICs of four districts).

Figure 8.3 exhibits the political map of the surveyed area, while the flow of collecting data is reflected in Fig. 8.4. Table 8.1 detects the block-wise number of samples collected, while the socio-demographic profile of the selected sample is exposed in Table 8.2.

8.3.3 Tools for Data Collection

The study has accessed 281 brickfield workers to collect the primary data through a structured questionnaire among the respective blocks (Joypur, Kotulpur, Hura,

[1] Where n = sample size, N = population (found as 1083), z = critical value of the normal distribution at the required confidence level (at 95% confidence interval 1.96), p = sample proportion taken as 0.05, e = margin of error taken as 5%. (https://www.wallstreetmojo.com/sample-size-formula)

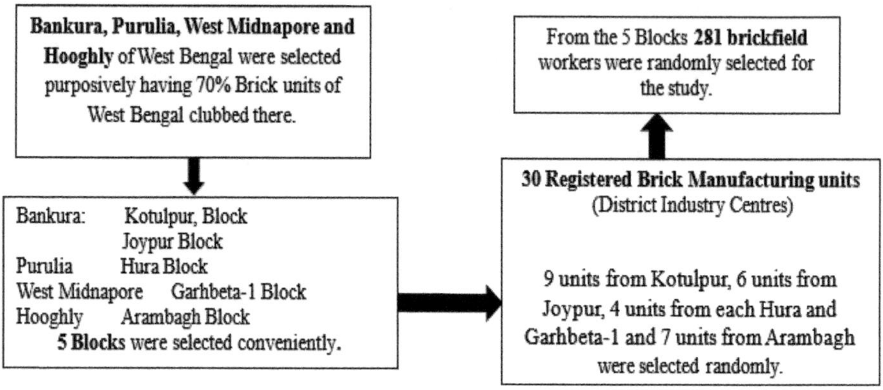

Fig. 8.4 Flow chart on the collection of the surveyed sample

Table 8.1 Block-wise sample collected

Particulars	Kotulpur	Joypur	Hura	Garhbeta-1	Arambagh
Brick units	9	6	4	4	7
Population	220	120	40	45	78
Sample	130	90	20	20	21
Total: 281					

Source Primary Survey (September 2023 to October 2023)

Table 8.2 Socio-demographic profile of the selected sample

Male	Female	Hindu	Others	SC	ST	Others	Illiterate	Primary	Up to 10th	10th +2
160	121	263	18	215	29	37	90	102	72	17
Total		281								

Source Primary Survey (September 2023 to October 2023)

Garhbeta-1 and Arambagh) by using printed hard copies during September–October 2023. Through the established scale of the National Institute for Occupational Safety and Health Worker Well-Being Questionnaire (NIOSH WellBQ 2021), a total of 18 items have been used in the field survey questionnaire to capture the perception of brickfield workers about workplace policy and culture, workplace safety measures, workplace health and hygiene practices and workers' health status. Subsequently, this established scale also provides an integrated assessment of workers' wellbeing across multiple spheres, including individuals' quality of working life, circumstances outside of work, and physical and mental health wellbeing. As many participants may not have an adequate literacy level, the study opts for unstructured interviews.

Here, mental health wellbeing was measured with items like 'problems in concentration', 'sadness' and 'work stress'. The assessment was in a negative sense as per the ILO-NIOSH-WellBO (https://www.cdc.gov/). In the same line, the work performance was also assessed in negative sense as per the standard questionnaire of ILO-NIOSH-WellBO, e.g. 'work in concentration' was measured in the negative sense like 'don't work in concentration' and so forth. This can be demonstrated in Tables 8.1 and 8.2, respectively.

8.3.4 Strategy of the Analysis

The study has applied the structural equation modeling (SEM) method to analyse the structural relationship between measured variables and latent constructs (Fan et al. 2016). Confirmatory factor analysis (CFA) is used to reduce the overall number of observed variables into latent factors based on commonalities within the data (Atkinson et al. 2011) and to evaluate the construct validity of the model. Subsequently, Path Analysis is used to find out the causal relationship among variables by creating a path diagram (Lleras 2005; Fan et al. 2016). Therefore, the study uses this approach to examine the consequences of the hypothesised model and the relationship between decent work-derived mental health and workers' work performance in two stages.

Initial Model: independent variables (decent work environmental components) and dependent variable (workers' mental health wellbeing).

Final Model: independent variables (decent work-derived mental health wellbeing) and dependent variable (workers' work performance).

In this regard, AMOS (Version 23) is used for structural equation modelling.

8.4 Results and Discussion

8.4.1 Reliability and Validity of the Scale

The study has gone through the reliability test of the scale, and the Cronbach's (alpha) value is 0.792, which confirms the reliability of the scale. Subsequently, Confirmatory Factor Analysis (CFA) is done to test the validity. In this study 18 observed variables are measured in 3 latent constructs like F_1: Workplace Policy and Culture (workplace respect, superior support, employer trust, job recognition, earning satisfaction, work values); F_2: Safety Measures (safety training, work resources, safety audit, safety equipment, improve safety, solve safety hazards) and F_3: Health and Hygiene Practices (safe drinking water, first aid, clean toilet and bathroom, clean basin and toilet soap, tiffin room, rest room). From Factor 1, two items (earning satisfaction and work values), Factor 2, three items (work resources, safety training and safety equipment), and from Factor 3, three items (first aid, tiffin room and rest room) were removed for the model fitness of the study. Hence, the study pursues 10 variables out of 18.

Table 8.3 Standardized regression weights: (default model)

Variables			Estimate
Job recognition	←	F1	0.677
Work value	←	F1	0.641
Work respect	←	F1	0.972
Employer trust	←	F1	0.749
Superior support	←	F1	0.960
Earning satisfaction	←	F1	−0.757
Solve safety hazard	←	F2	0.612
Improving safety	←	F2	0.676
Safety equipment	←	F2	0.338
Safety audits	←	F2	0.738
Work resource	←	F2	0.263
Safety training	←	F2	0.408
Restroom	←	F3	0.176
Tiffin room	←	F3	0.117
Handwash	←	F3	0.405
Toilet bathroom	←	F3	0.387
Frist aid	←	F3	0.783
Safe drinking water	←	F3	0.877

Source Author's calculation based on primary survey (September 2023 to October 2023)

8.4.2 Initial Model

Table 8.3 reflects the standardized regression weights: (default model) of the initial model, while the path diagram of the initial measurement model is shown in Fig. 8.5.

8.4.3 Final Model

Tables 8.4 and 8.5 respectively reflect the regression weights and standardised regression weights of the final model [independent variable—decent work-derived mental health wellbeing and dependent variable—workers' work performance], while Fig. 8.6 shows the final measurement model of decent work-derived mental health wellbeing and work performance.

8.4.4 Model Fitness

The overall model fit is assessed by using several fit indices (Gaskin and Lim 2016) measurement indicators. The model fit values are Comparative Fit Index (CFI) (a

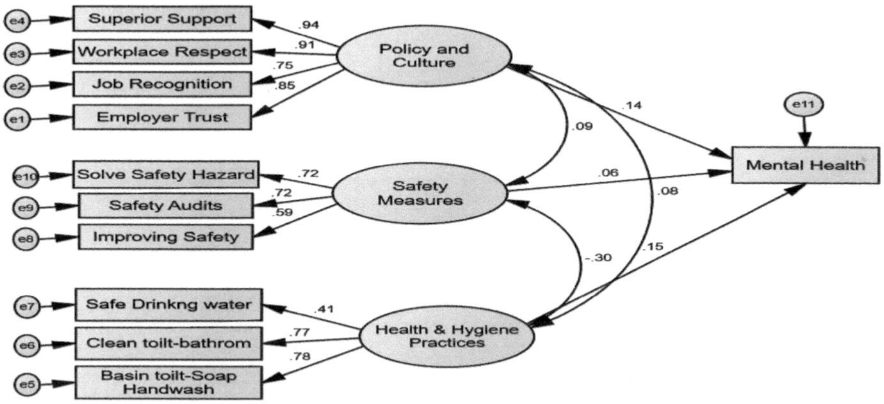

Fig. 8.5 Initial measurement model. *Source* Primary Survey (September 2023 to October 2023). Minimum was achieved Chi-square = 100.3 Degrees of freedom = 39 Probability level = 0.000

Table 8.4 Regression weights: (group number 1—default model)

Particulars			Estimate	S.E.	C.R.	P	Label
Mental health	←	F1	0.124	0.054	2.296	0.022	
Mental health	←	F2	0.081	0.100	0.809	0.419	
Mental health	←	F3	0.174	0.085	2.049	0.040	
F1Employertrust	←	F1	1.000				
F1Jobrecognition	←	F1	0.691	0.047	14.745	***	
F1Work respect	←	F1	1.161	0.057	20.349	***	
F1Superiorsupport	←	F1	1.202	0.056	21.510	***	
F2 improving safety	←	F2	1.000				
F2Safetyaudits	←	F2	1.159	0.156	7.431	***	
F2Solve safety hazard	←	F2	1.643	0.221	7.437	***	
F3Handwash	←	F3	1.000				
F3 clean toilet-bathroom	←	F3	1.199	0.170	7.063	***	
F3Safedrinking water	←	F3	0.635	0.112	5.680	***	
Work performance	←	Mental health	0.337	0.069	5.446	***	

Source Author's own calculation based on Primary Survey (September 2023 to October 2023)
Note Significance of Correlations: † $p < 0.100$; * $p < 0.050$; ** $p < 0.010$; *** $p < 0.001$

Table 8.5 Standardized regression weights: (Group number 1—default model)

Particulars			Estimate
Mental health	←	F1	0.141
Mental health	←	F2	0.060
Mental health	←	F3	0.152
F1Employertrust	←	F1	0.846
F1Jobrecognition	←	F1	0.746
F1Workrespect	←	F1	0.908
F1Supirior support	←	F1	0.943
F2Improvingsafety	←	F2	0.593
F2 safety audits	←	F2	0.724
F2Solve safety hazard	←	F2	0.720
F3 Handwash	←	F3	0.780
F3Clean toilet-bathroom	←	F3	0.771
F3Safedrinking water	←	F3	0.411
Work performance (WP)	←	Mental health (MH)	0.310

Source Author's calculation based on primary survey (September 2023 to October 2023)

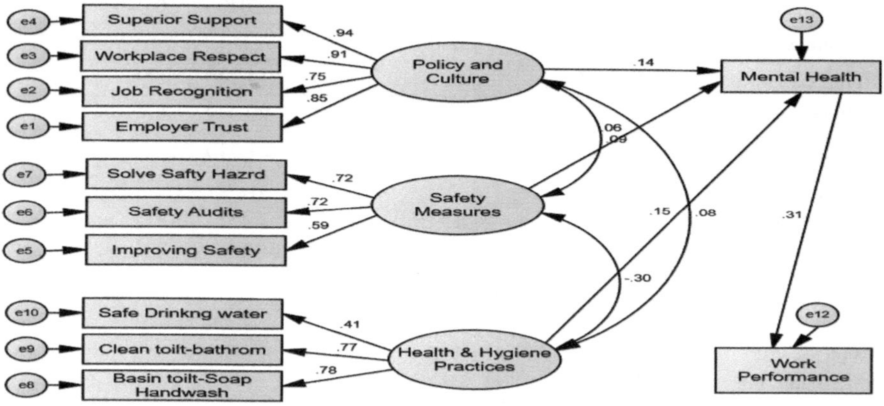

Fig. 8.6 Final measurement model of decent work-derived mental health wellbeing and work performance. *Source* Primary Survey (September 2023 to October 2023). Minimum was achieved Chi-square = 113.088 Degrees of freedom = 49 Probability level = 0.000

model comparison between the proposed model and a baseline model) = 0.951 (Excellent), root mean square error of approximation (RMSEA) (a quantification of error of the approximate fit) = 0.068 (Acceptable) (Goretzko et al. 2024). The standardised root mean square estimation (SRMR) is a 'badness of fit' measure as it quantifies the averaged square difference between each bivariate empirical correlation and the respective model-implied counterpart (Goretzko et al. 2024).

Table 8.6 Model fit measures

Measure	Estimate	Threshold	Interpretation
CMIN	113.088	–	–
DF	49	–	–
CMIN/DF	2.308	Between 1 and 3	Excellent
CFI	0.951	>0.95	Excellent
SRMR	0.062	<0.08	Excellent
RMSEA	0.068	<0.06	Acceptable
PClose	0.035	>0.05	Acceptable

Source Author's calculation based on primary survey (September 2023 to October 2023)

Hence, lower the value, better would be the fitness of the model. In the current study SRMR is found as 0.062 (Excellent). CMIN/DF (minimum discrepancy of confirmatory factor analysis/degrees of freedom) =113.88 which is between 1 and 3 (Excellent) and Probability of Close fit (PClose) is found as 0.035 (Acceptable). Thus, these values indicate that this structural model fits the data which is shown in Table 8.6.

8.4.5 Convergent and Discriminant Validity

The convergent validity of the study is established under three conditions as prescribed by Hair et al. (2009), (i) all individual items loaded significantly on their constructs, (ii) construct reliability (CR), an indicator of the internal consistency of a construct which is found to be greater than 0.7 and (iii) the average variances extracted (AVE) is representing the percentage of variances in a latent construct that should be greater than 0.5. In this study, the CR of the constructs ranged from 0.865 to 0.732 which indicates overall good internal consistency of the scale. The study also establishes discriminant validity as the correlations between different constructs are weaker than those within each construct. Here, in this construct of the study, AVEs are greater than r^2. Here the bold numbers (0.775, 0.591 and 0.591) represent the Square Root of the Average Variance Extracted for each latent construct (factor). The diagonal value of Factor 1 is 0.775, which is higher than its correlations with F2 (0.087) and F3 (0.084), whereas the diagonal value of Factor 2 is 0.591, which is greater than its correlation with F3 (-0.304**), and the diagonal value of Factor 3 is 0.591, which is also greater than its correlation with F2. Thus, the study satisfies the discriminant validity. Therefore, overall, the validity measurement indicators (Gaskin and Lim 2016) are suitable for further analysis and acceptable for the constructs in the present study, as shown in Table 8.7.

Tables 8.4 and 8.5 show that the value of coefficient regression between decent work-derived mental health wellbeing (MH) and work performance (WP) is 0.310 at probability value =0.000 while the standard estimated value is 0.69. In this

Table 8.7 Model validity measures

	CR	AVE	MSV	MaxR(H)	F1	F2	F3
F1	0.865	0.601	0.008	0.943	**0.775**		
F2	0.729	0.550	0.091	0.732	0.087	**0.591**	
F3	0.732	0.549	0.091	0.764	0.084	−0.304**	**0.591**

Source Author's calculation based on primary survey (September 2023 to October 2023)
Note CR: composite reliability; AVE: average variance extracted; MSV: maximum shared variance; MaxR(H): maximum reliability; F1: factor 1(policy and culture); F2: factor 2 (safety measures); F3: factor 3 (health and hygiene practice)

regard, though policy & culture and health & hygiene practice are found significantly related with mental health wellbeing of the workers, the safety measures cannot find any significant relationship with workers' mental health wellness as per their opinions.

The values show that workplace decent work-derived mental health wellbeing has a positive significant relationship with workers' work performance at workplace. Whereas Path analysis results indicated that a 1% change in decent work-derived mental health wellbeing generates a 31% change in work performance. This indicates that there is a positive relationship between mental health wellbeing and work performance.

Like other studies, the present study also is not free from some limitations. Here, the study suffers from subjective biases due to undue intervention of the employer. This was managed by the expertise of the interviewer there. Besides, the lower literacy level of the concerned participants was considered as a matter of problem-understanding in the right way. The sample measures may violate the universality of the study due to time and resource constraints though the size is supposed to be representative enough.

The study, hence, can be opted for other sectors and can also be carried to other dimensions of health-related wellbeing of individuals at the workplace like physical or combined physical and mental.

8.5 Discussion

Improving decent work conditions at workplace offers benefits to both employers and employees to get health wellbeing which resultantly helps to induce workforce performance, The previous literature (Mastroianni and Storberg-Walker 2014; Arb ogastet al. 2016; Abrampah et al. 2017; Alam et al. 2022a, b; On'gonge and Ngeno 2022; Monteiro and Joseph 2023; Mujtaba and Kaifi 2023; Johnson and Fujishiro 2023) supported the connectivity of decent work environmental components (workplace policy and culture, safety measures, health and hygiene practices) with health wellbeing. The present study detects that indecent work environmental components related to lack of standard work policy culture (workplace respect, superior support, employer trust, job recognition) and poor health and hygiene practices (safe drinking water, clean toilet and bathroom, clean basin and toilet soap and handwash) at the workplace are significantly responsible for the mental health illbeing (problems in decision making, sadness in work and work stress) of the brickfield workers which ultimately affects their performance. This was echoed in several previous findings also (Auten and Fritz 2019; Hennekam et al. 2021; Lu et al. 2022; Erwanto et al. 2024) where decent work is found directly related with workers' mental wellness.

8.6 Conclusion and Recommendation

The empirical study signifies the fact that the presence of a decent work environment at the workplace can contribute to the inclusion and improvement of mental health wellbeing of workers and have a direct mediating role to uplift workforce performance. This can resultantly minimise the psychological stress, sadness, absenteeism of workers and side-by-side can ensure retention and higher productivity of them. The positive horizontal and vertical inter-personal relationship along with standardised health-hygiene environment at workplace can ensure workers' mental health wellbeing. But, in this context, the brickfield workers are found lacking health hygiene awareness which frequently throws them to face extreme physical and mental trauma. This can be substantiated with the result of the present study where the responses of the workers do not reflect any significant relationship between safety measures, a component of decent work and their mental health wellbeing. The brickfield workers do not have proper training and skill to handle pressure and risk at work. Moreover, the sector suffers due to not having standardised policy to protect the health wellness of workers. All these are found to exaggerate the occupational hazards of the brickfield workers and thus threatens their mental health wellness and work performance at workplace.

Hence, to arrest the catastrophe, the employers of the sector must regularly analyse and monitor workplace safety measures with safety audits. The workplace hygiene practices with safe drinking water, clean toilets and bathrooms, clean basins and toilet soap, etc. are beneficial to the workers' health wellbeing (both physical and mental). The units should recruit occupational health-safety inspectors to take care of the maintenance of decent work environment and workers health connected therewith. Besides, more health-hygiene related awareness programmes should be organised at the workplace. This can ensure the positive intention of the unit to promote and protect workers' health and safety at the workplace and thus can improve their mental wellbeing and work performance.

The study may help in identifying, preventing and reducing indecent work environments associated with workers' health-related risks at the workplace. It provides proper insight into the major components of the workplace that are related to mental health wellbeing where mental wellness is found influencing workforce performance. Moreover, the study is based on the micro-empirical analysis which seems more effective in framing policy than that of any theoretical structure.

Hence, it proves its significance in front of policymakers, lawmakers, entrepreneurs and academicians.

Acknowledgments The authors express gratitude to those who participated in the survey for this study.

Declaration No potential conflict of interest is reported by the author(s) and no funding agency exists. The authors have no pertinent financial or non-financial interests to disclose.

References

Abrampah NM, Montgomery M, Baller A, Ndivo F, Gasasira A, Cooper C, Frescas R, Gordon B, Syed SB (2017) Improving water, sanitation and hygiene in health-care facilities, Liberia. Bull World Health Organ 95(7):526–530. https://doi.org/10.2471/BLT.16.175802

Alam M-U, Sharior F, Shoaib DM, Hasan M, Tabassum KF, Ferdous S, Hasan M, Rahman M, Tidwell JB, Zaqout M, Farah M, Rahman MA, Ahmed A, Ahmed T (2022a) Hygiene knowledge and practices and determinants of occupational safety among waste and sanitation workers in Bangladesh during the COVID-19 pandemic. Hyg Environ Health Wellbeing Adv 4:100022. https://doi.org/10.1016/j.heha.2022.100022

Alam S, Adnan ZH, Baten MA, Bag S (2022b) Assessing vulnerability of informal floating workers in Bangladesh before and during COVID-19 pandemic: a multi-method analysis. BIJ 29(5):1677–1702. https://doi.org/10.1108/BIJ-06-2021-0329

Arbogast JW, Moore-Schiltz L, Jarvis WR, Harpster-Hagen A, Hughes J, Parker A (2016) Impact of a comprehensive workplace hand hygiene program on employer health care insurance claims and costs, absenteeism, and employee perceptions and practices. J Occup Environ Med 58(6):e231–e240. https://doi.org/10.1097/JOM.0000000000000738

Atkinson TM, Rosenfeld BD, Sit L, Mendoza TR, Fruscione M, Lavene D, Shaw M, Li Y, Hay J, Cleeland CS, Scher HI, Breitbart WS, Basch E (2011) Using confirmatory factor analysis to evaluate construct validity of the brief pain inventory (BPI). J Pain Symptom Manage 41(3):558–565. https://doi.org/10.1016/j.jpainsymman.2010.05.008

Auten D, Fritz C (2019) Mental health wellbeing at work: how mindfulness aids in more ways than one. Organ Dyn 48(3):98–104. https://doi.org/10.1016/j.orgdyn.2018.04.001

Bajracharya SB, Gurung K, Mathema L, Sharma S, Mishra A (2021) Forgotten contributors in the brick sector in Nepal. Int J Environ Res Public Health 18(12):6479. https://doi.org/10.3390/ijerph18126479

Bansal N, Lanjewar A, Acharya S, Shukla S (2022) Assessment of pulmonary function test in brick factory workers. J Family Med Prim Care 11(11):6916. https://doi.org/10.4103/jfmpc.jfmpc_504_22

Bayram M (2019) Safety training and competence, employee participation and involvement, employee satisfaction, and safety performance: an empirical study on occupational health and safety management system implementing manufacturing firms. Alphanumeric J 7(2):Alphanumeric J2. https://doi.org/10.17093/alphanumeric.555154

Bentley T, Onnis LA, Vassiley A, Farr-Wharton B, Caponecchia C, Andrew C, Green N (2023) A systematic review of literature on occupational health and safety interventions for older workers. Ergonomics 66(12):1968–1983

Centers for Disease Control and Prevention Homepage. https://www.cdc.gov/

Chari R, Sauter SL, Petrun Sayers EL, Huang W, Fisher GG, Chang C-C (2022) Development of the national institute for occupational safety and health worker well-being questionnaire. J Occup Environ Med 64(8):707–717. https://doi.org/10.1097/JOM.0000000000002585

Chatterjee DS (2016) An innovative project under MGNREGS: making fly ash brick at Bankura District of West Bengal

Das B (2015) An evaluation of low back pain among female brick field workers of West Bengal, India. Environ Health Wellbeing Prev Med 20(5):360–368. https://doi.org/10.1007/s12199-015-0476-0

Das B (2020) Prevalence of work-related occupational injuries and its risk factors among brickfield workers in West Bengal. India Int J Ind Ergon 80:103052. https://doi.org/10.1016/j.ergon.2020.103052

Dery F, Bisung E (2022) Exploring informalities, inequalities, and health risk in the informal water sector in sub-Saharan Africa. Water Secur 17:100128. https://doi.org/10.1016/j.wasec.2022.100128

Dery F, Bisung E, Abudho B, Soliku O, Dery G (2023) Examining the health and wellbeing of women sanitation workers in Ghana and Kenya. Wellbeing Space Soc 5:100167

Erwanto E, Elshifa A, Wibowo TS, Yakin I, Salampessy M (2023) The influence of mental health on employee performance with physical activity as a moderating variable. Reslaj: Relig Educ Soc Laa Roiba J 6(3). https://doi.org/10.47467/reslaj.v6i3

Fan, Y., Chen, J., Shirkey, G., John, R., Wu, S. R., Park, H., & Shao, C. (2016). Applications of structural equation modeling (SEM) in ecological studies: An updated review. *Ecological Processes, 5*(1), 19.

Gaskin J, Lim J (2016) Model fit measures AMOS plugin. Gaskination's StatWiki

Ghosh B (2024) Assessing the effects of human interventions on the morphodynamics and health of a lowland tropical river: A case study of the river Dwarkeswar, India. Geosystems Geoenvironment 3(1):100234. https://doi.org/10.1016/j.geogeo.2023.100234

Goretzko D, Siemund K, Sterner P (2024) Evaluating model fit of measurement models in confirmatory factor analysis. Educ Psychol Meas 84(1):123–144. https://doi.org/10.1177/00131644231163813

Government of India: Ministry of Labour and Employment (n.d.) National policy on safety, health and environment at work place

Greiner BA, Leduc C, O'Brien C, Cresswell-Smith J, Rugulies R, Wahlbeck K, Abdulla K, Amann BL, Pashoja AC, Coppens E, Corcoran P, Maxwell M, Ross V, de Winter L, Arensman E, Aust B (2022) The effectiveness of organisational-level workplace mental health interventions on mental health and wellbeing in construction workers: a systematic review and recommended research agenda. PLoS One 17(11):e0277114

Hafee I, Yingjun Z, Hafeez S, Mansoor R, Rehman KU (2019) Impact of workplace environment on employee performance: mediating role of employee health. Bus Manage Educ 17(2):173–193. https://doi.org/10.3846/bme.2019.10379

Hair J (2009) Multivariate data analysis. Faculty and Research Publications. https://digitalcommons.kennesaw.edu/facpubs/2925

Hennekam S, Richard S, Grima F (2021) Coping with mental health wellbeing conditions at work and its impact on self-perceived job performance. Empl Relat: Int J 42(3):626–645. https://doi.org/10.1108/ER-05-2019-0211

Hessels J, Rietveld CA, van der Zwan P (2017) Self-employment and work-related stress: the mediating role of job control and job demand. J Bus Ventur 32(2):178–196. https://doi.org/10.1016/j.jbusvent.2016.10.007

Horváthová P, Mokrá K, Konečný F (2023) The level of occupational safety and health promotion in Czech family manufacturing enterprises. Saf Sci 157:105946

Home—Occupational Safety and Health—ILO Research Guides at International Labour Organization (n.d.) Retrieved 9 March 2024. https://libguides.ilo.org/occupational-safety-and-health-en

ILO (2015) Inspection L, Safety O, Morillons H (2015) Safety + Health for All (Safety and health at work). https://www.ilo.org/global/topics/safety-and-health-at-work/programmes-projects/safety-health-for-all/lang%2D%2Den/index.htm

ILO (1999) Decent work: report of the director-general. In: Report of the director-general. International labour conference. International Labour Office, Geneva

Johnson, Fujishiro (2023) Identifying occupational health inequities in the absence of suitable data: are there inequities in access to adequate bathrooms in U.S. workplaces?

Jones W, Gibb A, Haslam R, Dainty A (2019) Work-related ill-health in construction: the importance of scope, ownership and understanding. Saf Sci 120:538–550. https://doi.org/10.1016/j.ssci.2019.07.038

Kumari S (2018) Occupational health of brick workers of India. Int J Health Sci 9

Lleras (2005) Path analysis—an overview. ScienceDirect Topics. https://www.sciencedirect.com/topics/social-sciences/path-analysis

Lu X, Yu H, Shan B (2022) Relationship between employee mental health wellbeing and job performance: mediation role of innovative behavior and work engagement. Int J Environ Res Public Health 19(11):6599. https://doi.org/10.3390/ijerph19116599

Manimekalai N (2021) Women Workers in Unorganized Sector in the LPG context. Sanitation and Gender Concerns 3(7)

Manjunatha N (2023) Trishul division of mental health: conveying all sadness or stress of life is NOT a mental illness to people, the public, professionals, and policymakers. Indian J Psychiatry 65(9):983–984. https://doi.org/10.4103/indianjpsychiatry.indianjpsychiatry_286_23

Martins-Silva T, Hirschmann R, Bortolotto C, Fernandes M, Ruivo AC, Tovo-Rodrigues L (2020) Health self-perception and morbidities, and their relation with rural work in southern Brazil. Rural Remote Health. https://doi.org/10.22605/RRH5424

Mastroianni K, Storberg-Walker J (2014) Do work relationships matter? Characteristics of workplace interactions that enhance or detract from employee perceptions of Well-being and health behaviors. Health Psychol Behav Med 2(1):798–819. https://doi.org/10.1080/21642850.2014.933343

McCaughey D, DelliFraine JL, McGhan G, Bruning NS (2013) The negative effects of workplace injury and illness on workplace safety climate perceptions and health care worker outcomes. Saf Sci 51(1):138–147. https://doi.org/10.1016/j.ssci.2012.06.004

Mishra J, Acharya S, Taksande AB, Prasad R, Munjewar PK, Wanjari MB (2023) Occupational risks and chronic obstructive pulmonary disease in the Indian subcontinent: a critical review. Cureus. https://doi.org/10.7759/cureus.41149

Monteiro E, Joseph J (2023) A review on the impact of workplace culture on employee mental health wellbeing and well-being. Int J Case Stud Bus IT Educ (IJCSBE) 7(2). https://doi.org/10.47992/IJCSBE.2581.6942.0274

Mujtaba BG, Kaifi B (2023) Safety audit considerations for a healthy workplace that puts «people before profit» and OSHA compliance. Health Econ Manage Rev 4(1):11–25. https://doi.org/10.21272/hem.2023.1-02

National Brick Mission (2016) Chandra Bhushan, Dr D.D. Basu, Nivit Kumar Yadav and Rahul Kumar—A scoping paper, Centre for Science and Environment, New Delhi

Nanjunda DC, Venugopal PN (2022) Hard and invisible bricks in the wall: an empirical investigation on gender, caste, and health among migrant brick Workers in South India. J Anthropol Surv India 71(1):84–104. https://doi.org/10.1177/2277436X211046125

NIOSH Worker Well-Being Questionnaire (WellBQ) (2021) U.S. department of health and human services, public health service. Centers for Disease Control and Prevention, National Institute for Occupational Safety and Health. https://doi.org/10.26616/NIOSHPUB2021110revised52021

Obrenovic B, Jianguo D, Khudaykulov A, Khan MAS (2020) Work-family conflict impact on psychological safety and psychological Well-being: A job performance model. Front Psychol 11:475

Oakman J, Neupane S, Proper KI, Kinsman N, Nygård CH (2018) Workplace interventions to improve work ability: a systematic review and meta-analysis of their effectiveness. Scand J Work Environ Health:134–146

On'gonge HO, Ng'eno WK (2022) Influence of sanitation on employee performance at national social and security fund, Kenya. Int J Health Sci 5(3). https://doi.org/10.47941/ijhs.1126

Pandey, D. L. (2020). Work stress and employee performance: an assessment of impact of work stress

Rajan D (2022) Physical working environment and welfare facilities related factors causing heavy workload: an empirical study among sanitary workers. Lond J Soc Sci 2(3):19–52. https://doi.org/10.31039/ljss.2022.3.71

Saha MK, Ahmed SJ, Sheikh MAH, Mostafa MG (2020) Occupational and environmental health wellbeing hazards in brick kilns. J Air Pollut Health. https://doi.org/10.18502/japhv5i24242

Sari DL, Storyna H, Intan R, Sinaga P, Gunawan FE, Asrol M, Redi AANP (2021) The effect of job stress to employee performance: case study of manufacturing industry in Indonesia. IOP Conf Ser: Earth Environ Sci 794(1):012085. https://doi.org/10.1088/1755-1315/794/1/012085

Storbeck J, Stewart JL, Wylie J (2024) Sadness and fear, but not happiness, motivate inhibitory behaviour: the influence of discrete emotions on the executive function of inhibition. Cognit Emot. https://www.tandfonline.com/doi/full/10.1080/02699931.2024.2349281

Su X, Chan KL (2023) The associations of decent work with wellbeing and career capabilities: a meta-analysis. Front Psychol 14:1068599. https://doi.org/10.3389/fpsyg.2023.1068599

The World Bank (2007) The World Bank annual report 2007 [text/HTML]. World Bank. https://documents.worldbank.org/en/publication/documents-reports/documentdetail/732761468779449524/The-World-Bank-annual-report-2007

The World Bank (2020) World development report 2020: trading for development in the age of global value chains. https://www.worldbank.org/en/publication/wdr2020

Uddin M (2021) Addressing work-life balance challenges of working women during COVID-19 in Bangladesh. Int Soc Sci J 71(239–240):7–20. https://doi.org/10.1111/issj.12267

Webster E, Joynt K, Sefalafala T (2016) Informalization and decent work: Labour's challenge. Prog Dev Stud 16(2):203–218. https://doi.org/10.1177/1464993415623152

Willett JF, LaGree D, Shin H, Houston JB, Duffy M (2023) The role of leader communication in fostering respectful workplace culture and increasing employee engagement and Well-Being. Int J Bus Commun 23294884231195614. https://doi.org/10.1177/23294884231195614

World Health Organization (2010) Healthy workplaces: A model for action: for employers, workers, policy-makers and practitioners. Ambientes de TrabajoSaludables: Un Modelo Para La Acción: Para Empleadores, Trabajadores, AutoridadesNormativas y Profesionales. https://iris.who.int/handle/10665/44307

Manidipa DasGupta is attached with the department of Commerce of the University of Burdwan since 2013 as Assistant Professor. In her career as researcher, she is making her effort in publishing articles and book chapters in various national and international journals and Edited Books. Till now, she has completed four research projects from the University of Burdwan (Young Teachers' Research Project), ICSSR and NCW. Her research interest finds the areas like Entrepreneurship Development, Human Resource Development, etc. In recent times, she has been awarded ILO's SSTC scholarship to participate in the 8th Social and Solidarity Economy Academy at Seoul.

Sumanta Majumder is an Assistant Professor in the Department of Commerce at Bhatter College, Dantan (Autonomous), West Bengal. He is a research scholar at The University of Burdwan, West Bengal. His research interests focus on entrepreneurship development, human resource management, and other related areas, and his work has been published in various international and national journals, as well as Edited Books.

Investigating Transgender Inclusion in White-Collar Employment: Implications for Workplace Diversity Policies and the UN SDG Charter 2030

9

Vibhor Kataria and Tavleen Kaur

Abstract

Transgenders are a marginalized section of the society who are inappropriately clubbed in the LGBTQIA+ collective. Literature confirms that transgenders have their unique set of challenges that are much distinct from other members of the collective and yet they are brought together in this collective. By drafting a conceptual framework, this paper investigates the constructs to facilitate considerations of transgender inclusion in white-collar employment, in the specific context of early applicant or job-seeker stage. This conceptual paper also provides a foundation for an impending empirical investigation by drawing coherence by first establishing the rationale for considerations of transgenders as a unique entity that should not be a part of a collective. It then establishes the rationale for investigating transgender inclusion under the lens of societal and management considerations with the objective of extending workplace diversity policies taking into account considerations of recruitment practices at the early applicant or job seeker stage. The proposed research investigation also establishes the rationale for underpinning the research in the construct of the signaling theory. This paper is novel as it proposes a conceptual understanding for a novel empirical investigation of transgender inclusion in white-collar employment with implications for workplace diversity policies and considerations for UN SDGs 8 and 10 thereby impacting academia, society, and management practices.

V. Kataria (✉)
EFPM, Fortune Institute of International Business (FIIB), New Delhi, India
e-mail: vibhor.kataria-efpm@fiib.edu.in

T. Kaur
Fortune Institute of International Business (FIIB), New Delhi, India
e-mail: tavleen.kaur@fiib.edu.in

B. Kaur et al. (eds.), *Integrating Industry 5.0 and 4.0 for Technology and Humanity*, Contributions to Environmental Sciences & Innovative Business Technology,
https://doi.org/10.1007/978-981-92-1780-9_9

Keywords

Transgender • UN SGD • LGBTQIA+ • Workplace diversity • Inclusion

9.1 Introduction

The "T", within the LGBTQIA+ collective, representing transgenders is non-representative of the transgenders! Subsuming transgenders in the broader LGBTQIA+ collective is not representative of the unique challenges transgenders face (McFadden 2020a, b). Transgenderism is distinct from sexual orientation. This fundamental premise is the foundation of the proposed research. Despite the global acceptance of lesbians and gays, transgender rights remain unaddressed (Sawant 2017), and this presents a crisis (Jones 2020). The merging of transgender issues into the broader LGBTQIA+ concerns is misrepresentation of transgender-specific requirements (Salter and Sasso 2021; McFadden 2020a). Galupo et al. (2016) argue that the grouping of transgenders into the LGBTQIA+ collective fails to effectively address the unique challenges related to gender identity. Transgenderism is inherently about gender identity, and not sexual orientation. This is a compelling fundamental difference bringing to light the distinction of transgender challenges versus others in the collective. Mann (2021) highlights the legal requirements for surgical procedures to change gender markers on identification documents of transgenders, making the case for the distinct challenges transgenders encounter. Further, Johnson et al. (2021) discuss the naming at birth implications in light of the fact that future gender identity may not align with names assigned at birth. Such concerns are likely to exacerbate the challenges of workplace discrimination and lack of inclusivity of transgenders (Chaney and Hawley 2018; McFadden 2020a). Köllen and Rumens (2022) describe the workplace experiences of transgenders as being neglected by management researchers. Zachariah and Nandini (2023) and Drydakis (2025) emphasize this by highlighting the scarcity of research on transgender inclusivity in the workplace. Although research on LGBTQIA+ diversity is gaining prominence as the construct of DEI: Diversity, Equity, and Inclusion, with B (Belongingness) adding on to the narrative, attains vocalization, studies specifically about the transgender population remain limited (Panzia and Moresco 2022). Jones (2020) and Mor et al. (2025) advocate for organizational studies that specifically address transgender matters. As this research acknowledges the unique experiences of transgenders, focused organizational studies are necessitated. Literature discusses the requirement for the development of targeted strategies for enhancing inclusivity and addressing the specific challenges faced by transgenders at work as well as in the society.

9.2 Review of Literature

The criticality is well established in literature of the Person-Organization fit as the alignment between the values of the organization and those of potential employees. Person-Organization Fit (P-O Fit) is pivotal for not only the attraction but retention of talent as well (Cable and Judge 1996; Thomas and Wise 1999). We have discussed at large the nuances respective to the inclusion of transgenders grounded in the construct of the signaling theory by virtue of organizations signaling to potential job applicants the commitment to actively promote inclusivity within the culture of the organization. By sending out signals 'activating' and 'pointing' of inclusion, organizations are likely to enhance organizational attraction, which may, ultimately, lead to job pursuit intentions. Firms that cultivate diversity-focused climates are perceived more favorably by job seekers (Avery et al. 2013; Lambert 2015; Roberson et al. 2005; Thorsteinson et al. 2004). Specifics of workplace diversity policies if integrated into recruitment messages can potentially not only increase the overall attractiveness of the organization but also contribute to enriching the applicant pool by way of diversification (Bradley et al. 2023). Such integrated messaging signals to potential applicants that the organization values inclusivity, which can be particularly appealing to individuals from underrepresented groups. Research suggests that in order to enhance the impact of diversity and inclusivity more than generic affirmative action policies are required. Tailored and identity-conscious recruitment strategies need to be implemented, complemented via potential yields of activating and pointing signals of inclusion. Tailored approaches positively impact diversity outcomes for improved and effective organizational dynamics (Walker et al. 2007).

Acknowledging is the first step in addressing the unique challenges faced by transgender individuals, enabling organizations to become more equitable. To complement the recognized criticality of diversity and inclusion, this research has been able to establish several gaps in research to facilitate the economic inclusion and mainstreaming of transgenders at the workplace. While several of these gaps have been discussed in the below sections.

Most of the relevant research has clubbed transgender experiences under the broader LGBTQ+ umbrella, attributing the failure of distinctly establishing the impact of the empirical research on transgender individuals (Walker et al. 2007).

9.2.1 Transgender Inclusion Merits Consideration in Research for Organizational Practice

McFadden (2020a, b), Flores et al. (2016), Herman et al. (2022), and Wilson and Meyer (2021) highlighted that the younger generation is increasingly identifying with fluid and nonbinary gender identities. Such increasing expression of transgenderism makes a case to research, for organizational practice, work-related discrimination faced by these individuals. Literature thus makes the case for workplace diversity policy-making to address such emerging trends. Thoroughgood et al. (2020) and Abreu et al. (2023) highlight that despite such growing awareness, employers are not prepared to support transgender employees. The lack of significant organizational knowledge about transgender issues is confirmed by Hennekam and Köllen (2023). The authors suggest that without proper understanding of transgender issues, inclusivity efforts will be inadequate. There is thus advocacy for transgender-inclusive hiring practices in addition to transgender-affirming work conditions. These emergent trends thus necessitate the need to alter recruitment protocols in explicit support of transgender applicants through extended workplace diversity policies.

Zachariah and Nandini (2023) echo this criticality when they assert the timeliness of corporations to align with the emerging reality of impending count of transgender employees in the workforce. The research by Goryunova et al. (2021) is on the best practices to achieve a transgender-inclusive workplace, in which the authors provide valuable insights and strategies. Addressing the unique and distinct needs of transgender employees is not only a matter of social justice but also a critical factor in fostering productive and harmonious workplaces that would ultimately contribute to organizational productivity. Such a thought process will align with the imperative to develop targeted strategies that enhance inclusivity, ultimately benefiting employees as well as the corporations.

9.2.2 Transgender Considerations at the Early Applicant or Job-Seeker Stage is a Research Gap

There is an extensive discussion in literature of the workplace challenges experienced by transgender individuals as well as initiatives that corporations should take for their inclusion. There are descriptions of the workplace culture and the respective strategies employed within the broad LGBTQIA+ collective for navigating the workplace environment (Sabharwal et al. 2019; Fletcher and Marvell 2023; Jones 2020; Mor et al. 2025). Studies in literature predominantly focus on the experiences of LGBTQIA+ employees after they have entered the workforce, with discussions on how these employees manage identity disclosure and discrimination, in an effort to achieve inclusion for experiencing belongingness. While reviewing literature, the lack of representation of the experiences of transgender individuals during recruitment, particularly in the early application stage for transgender job seekers was noticeable. Abreu et al. (2023), in the context of their study, researching the challenges of immigrant Latinx transgender individuals during early recruitment stages, highlight the compounded barriers on account of the intersecting identities of gender and ethnicity that result in the marginalization of transgender job seekers leading to limited access to employment opportunities. On a similar note, Fasoli et al. (2024) have researched the discrimination of trans men in the hiring process. These two studies establish the case to research systemic biases that disadvantage transgender applicants even before they enter the workplace. Literature discusses at large the importance of the recruitment phase in terms of shaping the perception of job seekers' for subsequent engagement prospects with potential employers. Kleintop (2021) establishes that many transgender employees decide to "out" themselves at work depending on their perception of a psychologically safe environment that would enable them to be their authentic selves. Literature further clarifies the influence of the initial recruitment interactions on perceptions of job seekers on an organization's commitment to diversity and inclusion.

Resonating with an organization's culture and values is a strategic recruitment practice, the Person-Organization Fit, in the context of the marginalized community of transgender individuals. Thomas and Wise (1999) describe that alignment of recruitment strategies with the culture of an organization tends to draw individuals who not only possess the required skills but also mirror the organization's values. Such an alignment will trigger a perception of inclusion by way of fostering a sense of belonging that will reduce, in the long run, attrition on account of perceived misfit on account identity-related matters. Research has established Realistic Job Previews (RJPs) as an effective recruitment tool to trigger organizational attractiveness. As per Thorsteinson et al. (2004), RJPs provide a transparent and accurate assessment of not only the job role but also the environment in the organization. For transgender applicants, RJPs that explicitly communicate an organization's workplace diversity policies and practices can foster their perception of psychological safety, triggering job pursuit intentions via organizational attractiveness. Researching the job-seeking stage can establish specifics of discrimination experienced by transgender job seekers, such as the absence of inclusive language in job advertisements. These specifics can enable listing of specific signals that organizations can eventually use to enhance the job descriptions for triggering a stronger perception of inclusion versus conventional statements of diversity, equity, and inclusion. While significant progress has been made in understanding the workplace experiences of LGBTQIA+ individuals, and literature abounds with such investigations, research of literature presents a gap to research job-seeking experiences of transgenders, in specific, and at the early applicant or job-seeking stage. Effort of such a research will be a novel contribution to management research by way of contributing to a more inclusive organizational culture that will ensure that transgender job seekers receive equitable treatment from the start of the employment process.

9.2.3 Job Descriptions: A Tool to Draw Applicant Attention

Once the pivotal research gap of transgender inclusion at the early applicant or job seeker stage has been established, the logically coherent "next construct" is job descriptions (JDs). The imperative-criticality of job descriptions emerges as a tool to communicate signals of fostering perceptions of inclusion by way of attracting transgenders through shaping their perceptions of potential employers. Bradley et al. (2023) and Avery et al. (2013) highlight that job descriptions that carry signals of diversity and inclusion influence organizational attractiveness and the incumbent job pursuit intentions. One of the signals that can potentially influence these constructs is the use of transgender inclusive language. In fact, this belief is the trigger for the proposed research that was based on the premise that transgender inclusive language would trigger stronger feelings of attraction to transgender job applicants versus, in many cases, the potentially performative construct of conventional statements of diversity, equity, and inclusion. With this comparison forming the base or the foundation of the proposed research, the context of the research extends to discover, and describe activating and pointing signals of inclusion respective to the transgender job applicants thereby establishing for consideration of extension to workplace diversity policies, with implications for the UN SDGs 8 (Decent Work and Economic Growth) and 10 (Reduced Inequalities).

McFadden (2020a, b) and Wille and Derous (2017) point to the messaging within job descriptions as being critical for attracting diverse applicants and genuinely conveying an organization's inclusivity and diversity values thereby fostering a Person-Organization (PO) fit. Roberson et al. (2005) emphasize the significance of job descriptions as a potent tool that organizations can utilize effectively for providing details about job roles as well as the culture in the organization with the objective of enhancing perceptions of organizational attributes for a better person-organization fit. The wording of job descriptions can, potentially, either encourage or dissuade potential applicants, particularly those from underrepresented and marginalized groups such as transgenders. Clear and inclusive messaging presented in a compelling manner can mitigate concerns about potential discrimination and promote a sense of belonging for organizational attractiveness and the consequent job pursuit intention. In essence, thus, job descriptions that are able to relay details that can enable applicants to make an assessment of the match of their value aspirations with those of the organization are critical. This assessment of match is particularly relevant for transgenders to perceive being their authentic selves without fear of discrimination. As per Cable and Judge (1996) perceptions of person-organization fit influenced by job descriptions are critical during the recruitment process. Such perceptions shape job choice decisions and organizational entry with implications for employee retention.

We can thus effectively infer that when applicants, especially the ones who are marginalized, such as the transgenders, perceive a true and strong fit with the culture and values of an organization based on the signals of inclusion embedded in the job descriptions, the probability of the applicants to pursue employment opportunities is higher. Thus, it merits consideration for the organizations' HR and the business, to expend efforts in crafting job descriptions that reflect strongly on the culture of the organization, its values as well as its commitment to diversity, equity, inclusion, and belongingness. In his research, Lambert (2015) analyzes how job descriptions communicate both tangible and intangible job attributes, influencing organizational attractiveness and job pursuit intentions. The constructs of tangible and intangible job attributes, potentially connect respectively with the constructs of activating and pointing signals defined by Connelly et al. (2011) in their extension of the Signaling Theory of Spence (1971). Tangible attributes potentially include the items of salary, benefits, and job responsibilities, while intangible attributes encompass organizational culture and values. By effectively communicating the intangible attributes, job descriptions can significantly shape job seekers', especially the marginalized sections of the society, the transgenders, perceptions and intentions. Job descriptions can thus be considered powerful tools for influencing applicant behavior.

In line with the discussion about the tangible versus intangible attributes in the research work of Lambert (2015), Walker et al. (2007) suggest presentation of the discussion about the organizational policies in recruitment materials can influence organizational attraction and job pursuit intentions. To foster feelings or more aptly and firmly perceptions of inclusion particularly among the marginalized communities, job descriptions can play a critical role. Accordingly, it may be inferred from the review of literature that ambiguous or run-of-the-mill, generic job descriptions will fail to make an impact. While such a situation may not deter potential applicants from underrepresented groups, it definitely will not be a trigger. In contrast, job descriptions that are embedded with specific and clear messages, or signals, in the context of targeting underrepresented groups can enhance organizational attractiveness, and the consequent job pursuit intentions, by signaling an inclusive and supportive work environment.

9.2.4 Transgender Inclusion Under the Lens of the UN SDG 2030 is a Gap for Research

The premise of the United Nations Sustainable Development Goals (SDGs) as established by a review of literature is the aspiration to attain a future that is equitable, sustainable, and prosperous for all under the consideration of the following 3 dimensions: social, economic, and environmental. However, despite engagement at these three levels, or the extensive expanse of the considerations of the 17 UN SDGs, under the charter of the UN SDG 2030, the under-representation of the transgenders as well as their being overlooked in the discussions for achieving these UN SDGs is evident. Accordingly, this research has established the considerations of the transgenders in the charter of the UN SDG 2030 as a visible research gap. Additionally, the economic inclusion of the transgenders under the consideration of the UN SDG lens is an effective area of research that can make valuable and novel contributions to management research. Accordingly, the research proposal seeks to establish and address the gaps in the economic inclusion of the transgenders under the lens of the UN SDGs by focusing particularly on Goals 8 (Decent Work and Economic Growth) and 10 (Reduced Inequalities). A review of literature provides evidence of considerable discussion in the context of UN SDGs pertaining to Health (UN SDG 3); Industry, Innovation, Infrastructure (UN SDG 9); and Gender Equality (UN SDG 5), however, there, evidently and visibly, is limited focus in these discussions on the specific challenges, issues and concerns respective to transgenders. For example, although one paper discusses LGBT issues in the context of SDG 5 (Gender Equality), the descriptions are in the context of women and girls, subsuming cis-normative versus trans-specific nuances. Such a stance is a reflection on the potentially limited understanding of the nuances associated with transgender identity, thereby limiting the discussion to cis-normative terms. Aggleton et al. (2023) critique the absence of discussions related to gender-diverse individuals in the SDGs. They further note the incumbent social and economic discrimination of transgender individuals. Such discrimination has potential implications on employment opportunities for transgender individuals, which may be construed as a key concern by virtue of the scope of UN SDGs 8 and 10. Noting the recent studies in literature, specifically, by Drydakis (2025); Hutchinson et al. (2024); Walia and Mondal (2025), the pervasive discrimination against trans and nonbinary workers, contributing to high rates of unemployment, underemployment, and economic precarity is highlighted. These observations have been based on the scope of the UN SDG 8 that references decent work and economic growth. We can infer from the discussion in literature of the importance of workplace diversity policies facilitated by empirical research that would contribute to advancing employment rights of transgender individuals with implications for achieving the fairness and equality goals of SDG 10 (Reducing Inequalities). Based on the discussion by Gore (2025) of governance that equally distributes social, economic, and environmental goals, we may infer the criticality of considerations in the UN SDGs for inclusion of transgenders. Additionally, by noting the advocacy by Riha et al. (2025) for the integration of gender issues as a theme that cuts across all of the 17 UN SDGs, we may infer a policy gap that calls for economic stability and reducing inequalities by fostering gender-inclusive, extending beyond the binary, workplace environments. In conclusion of the discussion of the UN SDGs in the context of transgender inclusion we may propose the novelty of the research that aims to systematically explore the principles and scope of SDG 8 and SDG 10 to extend the integration of the specific needs and challenges of the transgender community into the framework with potential impact on global and national policies for fostering diversity, equity, and inclusion.

9.2.5 Methodological Gap

In literature, positivist methods are critiqued on account of their limitations in describing human behavior (Crossan 2003). While quantitative studies are effective in providing valuable statistical insights, they cannot deliver on the precise personal narratives of lived realities. Positivism is anchored in the fact that reality is measured and then interpreted via objective observation and empirical analysis. However, Crossan (2003) challenges this perspective on account of the limitations of quantitative studies in establishing the meaning that beings assign to their individual, qualitative experiences. With respect to researching transgender inclusion, a quantitative research methodology will not be effective considering that it overlooks the requirement of understanding and establishing the deeply personal complexities of transgender concerns and challenges that potentially impact the job-seeking processes of transgenders. For example, statistical insights can establish the percentage of transgenders employed in various sectors of the economy as well as establish the respective factors impacting the inclusion of the percentage of transgenders in the respective sectors; however, what causes these factors to be the underlying impetus for transgender inclusion in these sectors requires qualitative insights.

We may thus infer and conclude to establish the methodology of the proposed research work to be qualitative based on the constructivist or interpretivist research paradigm. Considering the complexities associated with the elements of identity, discrimination, and perceptions, a nuanced exploration is required versus what a positivist research paradigm can provide. Given the limitations, as discussed, a qualitative research approach is thus rationalized for the study in context. To reiterate, a qualitative mode of inquiry is better equipped to handle an insightful exploration to capture and narrate the emotions and interactions based on the narratives of an individual, thus providing an insightful exploration of the phenomena under investigation.

The LGBTQIA+ collective overlooks the unique challenges and experiences of transgenders, revealing a gap, concurrent with providing a rationale for focused research on transgender inclusion in the context of the distinct challenges of transgenders.

Review of literature revealed a critical gap by highlighting the treatment in most research of transgender discrimination within the workplace. Research on the early applicant or job seeker stage was not visible. This is critical considering the many discussions in the earlier sections that perceptions of inclusion, particularly in the context of under-represented groups that are much sensitive to unconscious bias, let alone discrimination, are established from the signals communicated in job descriptions prompting the acts of organizational attraction and job pursuit intentions. Research focused on workplace discrimination very evidently overlooks discrimination even before the commencement of employment, i.e., during the job-seeking and-or early applicant stage. Furthermore, a dominance of quantitative methodologies in transgender empirical studies was observed. While quantitative research can provide valuable insights and generalizability, it cannot capture the lived experiences of transgenders that are critical for discussion related to workplace diversity policies as well as implications for policymaking in the scope of the UN SDG agenda.

Literature highlights the crucial role of inclusivity and diversity in enhancing organizational attractiveness and effectiveness. However, significant gaps are identified, particularly concerning the unique experiences of transgender individuals in the recruitment process. By the review of literature, a clear need for focused, qualitative research that explores the experiences of transgenders during the early applicant stage is well established for impact on workplace diversity policies respective to the scope of the agenda of UN SDGs 8 and 10.

9.3 Theoretical Underpinning

Signaling theory (Spence 1978) as propounded by Spence establishes a framework that enables researchers to analyze interactions between the sender and the receiver. Such interactions occur under conditions established of information asymmetry, as established by the theory. The theory describes information asymmetry to occur when one party (the job seeker) is in possession of information that is less than the information possessed by the other party (the employer). Drawing cues from this fundamental premise as established by the Signaling Theory, we can, thus, infer that in the context of human resource management, signaling can be key to how organizations project their culture and the respective associated values to potential employees. Such signals can thereby influence conditions of organizational attraction as well as job pursuit intentions. To establish further the theoretical underpinning of the Signaling Theory as in the context of the proposed research and the incumbent conceptual framework for the research, let us understand with a discussion of the following core concepts of Signaling Theory as may be relevant to the impact of the proposed research on the practices of HR Management.

9.3.1 Information Asymmetry and Organizational Signaling

Organizations need to be strategic in their use of signals in the management of the discussed-described information asymmetry. By showcasing with a description of qualities like workplace inclusiveness, quality of employee benefits, and corporate culture as well as other relevant signals, which are potentially pivotal in attracting prospective employees, organizations can leverage on the existing information asymmetry between themselves and job applicants to influence application organizational attraction and job pursuit intentions. Playing on these signals to even out the information asymmetry is potentially crucial to facilitate the organizational attraction and job pursuit intentions of transgender job seekers who rely on such cues to gauge an organization's commitment to inclusivity, a parameter that will influence their job application decisions.

9.3.2 Quality and Honesty of Signals

Connelly et al. (2011) worked on the seminal work of Spence (1978) to extend the discussion of Signaling Theory by establishing that effective organizational signals, the activating signals will be costly, observable, and honest. These costly, activating signals indicate a genuine investment of the organization in the policies advertised, such as transgender-specific health benefits thereby impacting employer credibility on the consideration of diversity, equity, and inclusion. Activating signals versus pointing signals can be constructed with the parameter of honesty. Honest signals are those that accurately reflect the organization's true practices, aligning applicant expectations with reality. In effect, honesty and cost are mutually inclusive. A costly signal would by default be honest and are referred to as "activating signals." On the flip side, "pointing signals" may not be dishonest, but could be performative, or as an item in the checklist, referred to as "performative allyship" as in the construct of signals of inclusion in the context of transgender job applicants.

9.3.3 Applicability of the Signaling Theory to Research on Transgender Inclusion

Literature has established the unique challenges confronted by transgender job seekers in the context of the uniqueness of their gender identity. They can thus be particularly sensitive to the activating versus pointing signals integrated into job descriptions by potential employers regarding their commitment to diversity and inclusion, in the specific context of the marginalization of the transgender community. The proposed empirical research focuses on the signaling mechanisms organizations can use to communicate their inclusivity toward transgenders specific to white-collar employment. The uniqueness and novelty of the proposed research, based on identified gaps from the literature, are incumbent on the research being exploratory in terms of establishing via qualitative, content, and thematic analysis, the specific reasons for signals (activating) that trigger a perception of inclusion while others are irrelevant or passive (pointing signals). Additionally, the research dominates by virtue of it empirically investigating these factors or signals of inclusion at the early-applicant stage versus discrimination or challenges that are replete in research specifically in the context of the workplace.

Cheng and Roy (2023) note that organizations with more female directors are more likely to implement comprehensive trans-inclusive health policies. Offering trans-inclusive health benefits logically affect the well-being of transgender employees and could serve as a critical factor in their perception to concede a welcoming environment seeded in diversity, equity, and inclusion measures. On the flip side, integrating trans-inclusive language into corporate communications and imagery in marketing materials, while symbolic or representative of diversity and inclusion, involves minimal costs in comparison to organizations affording comprehensive trans-inclusive health insurance packages. While the costs may be lower comparatively, a thoughtful approach to ensure authenticity of trans inclusion is representative of the effort expended. Such signals, while less credible and honest, could potentially play a role in setting an inclusive tone, making potential transgender employees develop a sense of inclusion from the outset. Ladwig (2023) discusses the low-cost, high-impact of leadership style that fosters inclusivity along with the careful use of language in internal and external corporate communications. Such approaches to diversity, equity, and inclusion over and above high-cost, high-impact interventions are likely to influence the climate of the organization, thereby presenting to current and potential employees a psychologically safe environment by supporting diversity and inclusion.

The research of Majid et al. (2023) presents quantifiable indicators the impact of which can be explored as workplace signals on transgender inclusion. The findings of the researchers showcase the statistical impact of indicators (signals) on job satisfaction, retention rates, and overall workplace harmony. Identification to establish such signals through primary and secondary research will be utilized to build the assertion of this research. The research asserts the strategic identification and utilization of both activating and pointing signals to effectively communicate their commitment to transgender inclusion and establish the impact of these on perceptions of inclusion among the transgenders. Next would be to consider the trigger for the thought of the current, proposed research, it would be wise to discuss the impact-implications, in the next section, of the transgender-inclusive language in facilitating perceptions of inclusion versus conventional statements of inclusion.

9.3.4 Signaling Inclusion by Transgender Inclusive Language

The integration of transgender-inclusive language perhaps could be a potent form of signaling inclusion, active or pointing, as per the constructs set forth by the signaling theory if one were to make a comparison with conventional statements of inclusion often included by most organizations that perhaps are performative rather than active. In literature, criticality of language use in shaping workplace culture and individual identity affirmation is discussed. Ladwig (2023) has presented practical recommendations on how language use can enhance transgender, gender diverse (TGD), and broader LGBTQ+ inclusion within professional settings. These recommendations highlight the use of carefully chosen transgender-inclusive vocabulary that acknowledges and respects the diverse gender identities of employees. Further, Zimman (2017) extends the role of language beyond affirmation of individual gender identity to the institutional implications of language by way of asserting the impact of the construct of language in organization policies as well as in corporate communication, impacting perceptions of an inclusive work environment. This perspective is crucial for understanding and establishing the potential impact of language, as well as extending to other signals, on triggering perceptions of workplace inclusion among transgender applicants. Gaucher et al. (2011) additionally provide evidence of the impact of gender-neutral language in recruitment materials. Their studies suggest that gender-neutral terminology minimizes unconscious bias and thus, potentially broadens the appeal of job postings to a diverse pool of applicants. Such initiatives are thus likely to improve the overall diversity in the organization.

9.4 Finding and Discussion

The research provides recommendations for corporations attempting to enhance inclusivity as well as support for transgender employees. The authors thus highlight practical steps for implementation of the economic inclusion of transgender employees with their acceptance in the workplace. The nuanced challenges and the incumbent experiences of transgender applicants and employees necessitate studies to establish a focus for their inclusion with practical interventions. Literature abounds with discussions on recruitment strategies that emphasize the importance of early interactions between applicants and organizations. For example: Rynes and Barber (1990), Thomas and Wise (1999), Ehrhart and Ziegert (2005), and Thorsteinson (2004) highlight in their discussions shaping of candidate interest to yield perceptions of organizational attractiveness and fit.

Positive encounters during the recruitment stage can enhance a candidate's enthusiasm based on the perceptions of alignment of the candidate and the organization's culture and values. From the discussion to this point, we conclude that despite the insights, consideration at the recruitment and early application stages for transgender job seekers remains underexplored in academic research. This is thus a gap that merits consideration for research to foster perceptions of inclusion and the incumbent fitment of transgender applicants into the organization. Addressing this gap is critical for extending workplace diversity policies to support transgender individuals from the outset of their employment journey. The use of the construct "job descriptions" can help overcome the underrepresentation and challenges faced by transgender applicants. In conclusion, the messaging of job descriptions is, potentially, critical for shaping perceptions about potential employers. The messaging triggers signals, drawing applicant attention and thus playing a role in the initial stages of recruitment, influencing applicants' decisions to engage with an organization. By crafting job descriptions with emphasis on the organizational values of diversity, equity, and inclusion, organizations can signal perceptions of inclusion, thereby attracting a range of candidates, including those from underrepresented groups such as transgenders. Addressing this aspect of signaling inclusion is essential for developing comprehensive strategies to extend workplace diversity policies that signal fairness and inclusivity right at the start of the employment journey. This brings us to thus establish the theoretical underpinning of this proposed research.

Based on the premise of the impact or potential of the various signals to bridge information asymmetry, Signaling Theory provides an effective and applicable theoretical framework for not only understanding but also building the constructs of the Signaling Theory for organizations to establish via the medium of job descriptions messaging by integrating signals to communicate their commitment to transgender inclusion. For example: Bradley (2023) establishes the distinction between activating and pointing signals of inclusion. Activating signals, as can be inferred based on the draw of literature, are tangible, honest and potentially costly, showcasing an organization's commitment to inclusion, requiring substantial investment and effort, hence activating signals are viewed as more credible and effective. On the other hand, pointing signals, while less resource-intensive, play a crucial role in shaping perceptions through symbolic actions and representations, but lack the credibility and effectiveness of the activating signals. By analyzing the gaps after establishing the details of the current implementations of SDGs 8 and 10 with implications for proposing the frameworks respective to the economic inclusion of the transgenders, this research aims to make a societal and practical impact by triggering food for thought to enable the formulation of more equitable workplace diversity policies that acknowledge and address the disparities faced by transgenders by mitigating potential unconscious biases at the early applicant stage. The potential oversight of transgender issues in the framework of the Sustainable Development Goals represents not only a gap in policymaking but also a missed opportunity for equality to equity-led economic growth. By focusing on the oversight-triggered gap in the UN SDG agenda, this proposed research seeks to advance the understanding of inclusive practices and promote a more equitable distribution in pursuit of sustainable and responsible business practices in support of the United Nations SDG agenda.

In the context of transgender inclusion in white-collar employment, a qualitative research methodology will be better and more effectively aligned to uncover and thus establish the subjective experiences of transgender job applicants seeking white-collar employment. With this conceptual exploration, the proposed research is structured to establish perceptions of person-organization fit based on organizational signals of inclusions integrated into job descriptions. Such an approach thus effectively aligns with the interpretivist or constructivist epistemology, which posits that reality is socially constructed and is best understood based on the subjective experiences of individuals.

In summary, applying a qualitative framework is expected to ontologically effectively fill the established gap in literature by way of establishing a comprehensive understanding of the enablers as well as barriers to transgender inclusion even before they enter the workforce. In foresight, a qualitative methodological approach is not only appropriate but imperative for this proposed research work. By aligning with the goal of capturing the nuances of transgender experiences, the proposed research methodology adds novelty to the research work attempting to study the phenomena related to the lived complexities of transgenders. The research is being conceptualized to establish a rich and detailed understanding of transgender inclusion in white-collar employment at the early applicant stage, further in-depth interviews can be conducted using thematic analysis. This research approach is thus expected to generate valuable insights that will extend beyond academic contributions, offering practical guidance to organizations as well as policy makers aiming to enhance efforts of diversity, equity, inclusion, and belongingness. Ultimately, the study will deliver on the primary objective of fostering environments where transgender individuals can perceive inclusion and thereafter thrive professionally.

Addressing the established gaps will not only contribute to academic understanding, respective to the Signaling Theory, but also support organizations in developing more inclusive practices, ultimately aligning with broader societal goals of equality and well-being respective to the construct of policy making by way of implications for the agenda of UN SDGs 8 and 10.

In future, in-depth understanding of the live experiences of the transgenders will provide insights into the subjective realities of transgender individuals with a focus on the dimensions of their job-seeking experiences. The research from the outset was conceptualized with the objective of establishing insights to guide policymakers and organizations in extending diversity policies for the economic inclusion of transgenders. The study focuses on transgenders seeking white-collar employment. This delimitation excludes other marginalized groups of the LGBTQIA+ collective and non-white-collar employment contexts.

Additionally, we expect the research to be focused on the instances of transwomen versus transmen. The sample size may potentially be limited on account of accessing a community that is not only marginalized but is also reclusive at the same time, thus presenting an extremely sensitive demographic to research with. Literature defines the self-reporting bias and the incumbent non-replicability inherent in any qualitative research. Data collected using the proposed approach of semi-structured interviews is subject to the self-reporting bias of the researcher and non-replicability on account of the subjective interpretations and potential biases of the researcher, which could influence the analysis for firmed-up findings. As is apparent by review of literature, the majority of the research has been in the context of workplace experiences. Research that does discuss hiring discrimination is limited with none of the specific instances of consideration of transgender inclusion at the early applicant or the job seeker stage in the specific context of the effectiveness of job descriptions. The research can be framed within the context of relevant UN SDGs 8 and 10, potentially limiting the exploration of relevant issues in the context of other UN SDGs that may also impact considerations of the social and economic inclusion of transgenders.

9.4.1 Formulation of Research Themes and Questions and Objectives

Based on the discussions in the preceding sections, the literature review and the compiled research gaps as well as the intent of the conceptualized research, the primary themes are: understand and establish the unique challenges of transgenders; understand and establish the lived realities of transgenders; understand and establish transgender discrimination at the early applicant or job seeker stage; establish and understand the Impact of early Recruitment Practices on Perceptions of Inclusion influenced or by way of the P-O Fit; establish and understand amongst the transgenders the Awareness and Effectiveness of Workplace Diversity Policies to thus in effect suggest extensions to existing workplace diversity policies by encompassing measures to mitigate unconscious bias against the transgenders as well as implications for the agenda of the UN SDGs 8 and 10 for a more comprehensive and inclusive charter. With the above 5 themes in context, the following are the proposed research questions and research objectives that the proposed research is conceptualized to address as mentioned in Table 9.1.

Table 9.1 Conceptualization of research questions and research objectives

Research Question-1: What are the challenges and discriminatory practices that are unique to the transgender experience in the early applicant or the job seeker stage in the Indian context in specific consideration of job descriptions as an effective tool to attract transgender applicants?	**Research Objective-1**: To sensitize organizations with potential unconscious bias resulting in early applicant stage transgender discriminatory practices emanating from messaging of job descriptions
Research Question-2: How do recruitment practices in the specific context of the job descriptions at the early stage job applicant or job seeker stage signal to the transgenders perceptions of person-organization fit?	**Research Objective-2:** To analyze the impact on perceptions of the person-organization fit of recruitment practices in the specific context of job descriptions at the early applicant or job seeker stage
Research Question-3: What are the social lived realities of transgenders in India, having white-collar aspirations in the reference to the early applicant stage?	**Research Objective-3**: To extend workplace diversity policies as well as establish implications for the UN SDGs 8 and 10 for economic inclusion of transgenders by way of understanding the social and professional experiences, in the context of the early applicant stage, of transgenders aspiring white-collar employment in India
Research Question-4: What is the perception of transgenders in the context of their inclusion within the broader LGBTQIA+ collective, given their unique challenges compared to the other groups of the collective?	**Research Objective-4**: To explore the transgender perceptions of inclusion within the LGBTQIA+ community, considering their respective unique challenges versus other members of the LGBTQIA+ collective
Research Question-5: Are transgenders cognizant of workplace diversity policies that posit unconscious bias against transgender inclusion, and thus, accordingly, which organizations are aspirational for them?	**Research Objective-5**: To assess transgender awareness of the statutes within workplace diversity policies to list organizations that are potential role-models for transgender employment in white collar jobs

9.5 Implications and Conclusion

Promoting economic inclusivity of the transgenders lends thrust to the proposed research in context and thus stands as the core of this study. By virtue of being able to identify and establish using qualitative insights the optimization of signals of transgender inclusion, the research lays a foundation for organizations as well as policymakers to work out initiatives for fostering perceptions of inclusion as well as sense of safety and security by triggering efforts to break down the challenges faced by transgenders. This measure thus will drive the development of a more inclusive economy and societal inclusion of transgenders impacting considerations for policymakers to drive a positive change in society in reference to potentially weeding out the marginalization of the transgenders over time. Also, a fundamental premise of the research is to impact workplace diversity policies, which at this time speak in most cases about the inclusion of the LGBTQIA+ collective, not so much about transgenders, at the workplace. The implications of the proposed research are targeted to sharpen management practices by refining workplace policies for transgender inclusion, specifically at the job entry stage versus the extensive discussion in literature with respect to the workplace. The presentation of the findings of the research will offer insights to guide managers for firming recruitment strategies by drafting and integrating messages for transgender inclusion, but before that, add to the rising awareness for transgender inclusion. With this, the research will also contribute to the scholarship of employer branding. Research indicates that signals of diversity and inclusion strengthen the employer brand, potentially attracting cis employees thereby enriching the applicant pool for organizations to choose from. Integration of early-stage recruitment practices to establish inclusive policies will enable organizations to build a comprehensive framework that will positively impact the strategic approach of organizations to diversity and inclusion.

As part of its theoretical underpinning, the current research aspires to extend on the future directions as laid out by the work of Connelly et al. (2011). Connelly et al. (2011) proposed that researchers expand signaling theory in new and complex ways. Researching signals that foster perceptions of inclusion for transgenders lends novelty to the treatment of the Signaling Theory, providing a potential extension with the consideration of the novel treatment of transgender inclusion in white-collar employment. We have been able to effectively interpret literature highlighting gaps in research for the treatment of transgenders as well as potentially making the business (ref: workplace diversity policies) and the societal (ref: UN SDGs 8 and 10) case for pursuing the proposed research. By researching qualitative reactions of transgenders to these signals, the research could potentially establish measures for organizations to establish cues to show their commitment to transgender inclusivity, enabling extension of workplace diversity policies. This research thus has the potential to impact academia, management practices, as well as societal policies alike. By delving into the nuances of signaling theory we can establish measures of transgender inclusion and thus empower organizations to become more welcoming by fostering among transgenders perceptions of supportive corporate environments.

This study thus will not only follow to build on the suggestions by Connelly et al. (2011) for academic extensions to the signaling theory but will also meet a critical need of inclusive business practices as well as societal inclusion of a marginalized section of the society. The established objectives of this research aim to establish elaborate insights into the unique challenges of transgenders with respect to their economic as well as societal inclusion in to the UN SDG Charter 2030. By focusing on insights into the challenges of the transgenders, our study will connect with researching the broader societal objectives as mandated by the UN SDG Charter 2030, captured in the powerful aspiration of: "no one left behind." The study explored 5 themes, 5 research questions and related objectives, which can be explored further to reduce the identified gaps. A key aspect that lends novelty to the research thus is to establish and develop an alignment by researching the nuances of the United Nations Sustainable Development Goals (SDGs) 8 (promoting inclusive economic growth) and 10 (reducing inequalities) for the social and economic policy making relevant to the respective charters of the UN SDG and connecting them with qualitative findings of the research to establish impact on policy.

References

Abreu RL, Gonzalez KA, Lindley L, Capielo Rosario C, Lockett GM, Teran M (2023) "Why can't I have the office jobs?": immigrant Latinx transgender peoples' experiences with seeking employment. J Career Dev 50(1):20–36

Aggleton P, Sciortino R, Newman CE (2023) Engaging with omission: promoting concern for gender and sexuality diverse people in SDG 5 and beyond. In: Transforming unequal gender relations in India and beyond: an intersectional perspective on challenges and opportunities. Springer, Singapore, pp 315–326

Avery DR, Volpone SD, Stewart RW, Luksyte A, Hernandez M, McKay PF, Hebl MMR (2013) Examining the draw of diversity: how diversity climate perceptions affect job-pursuit intentions. Hum Resour Manag 52(2):175–193

Bradley C, Moergen KJ, Roumpi D, Simon LS (2023) Don't just tell me, show me: impacting perceptions of organizational attraction and fit using activating LGBT diversity signals. Pers Psychol 77:1025

Cable DM, Judge TA (1996) Person–organization fit, job choice decisions, and organizational entry. Organ Behav Hum Decis Process 67(3):294–311

Chaney MP, Hawley L (2018) Developing human resource development competencies to manage sexual orientation and transgender diversity issues in the workforce. In: Diversity in the workforce. Routledge, pp 120–152

Cheng JYJ, Roy D (2023) Towards trans-inclusive workplaces: when do female directors matter? In: Academy of management proceedings. Academy of management, Briarcliff Manor, NY, vol 2023, no 1, p 16500

Connelly BL, Certo ST, Ireland RD, Reutzel CR (2011) Signaling theory: a review and assessment. J Manag 37(1):39–67

Crossan F (2003) Research philosophy: towards an understanding. Nurse Res 11(1):46–55

Drydakis N (2025) Employment discrimination against transgender women in England. Int J Manpow 46(1):58–74

Ehrhart KH, Ziegert JC (2005) Why are individuals attracted to organizations? J Manag 31(6):901–919

Fasoli F, Frost DM, Serdet H (2024) How voice transition and gender identity disclosure shape perceptions of trans men in the hiring process. Gender Work Org 31(1):36–58

Fletcher L, Marvell R (2023) Furthering transgender inclusion in the workplace: advancing a new model of allyship intentions and perceptions. Int J Hum Resour Manag 34(9):1726–1756

Flores AR, Brown TN, Herman J (2016) Race and ethnicity of adults who identify as transgender in the United States

Gaucher D, Friesen J, Kay AC (2011) Evidence that gendered wording in job advertisements exists and sustains gender inequality. J Pers Soc Psychol 101(1):109

Galupo MP, Henise SB, Mercer NL (2016) "The labels don't work very well": transgender individuals' conceptualizations of sexual orientation and sexual identity. Int J Transgender 17(2):93–104

Goryunova E, Schwartz AK, Turesky EF (2021) Exploring workplace experiences of transgender individuals in the USA. Gender Manag Int J 37(6):732–750

Gore E (2025) Towards a trans-inclusive critical international political economy? Or why trans oppression matters for understanding capitalism and social reproduction. Cap Class 49(1):123–144
Hennekam S, Köllen T (2023) Trapped in cisnormative and binarist gendered constraints at work? How HR managers react to and manage gender transitions over time. Int J Hum Resour Manag:1–27
Herman JL, Flores AR, O'Neill KK (2022) How many adults and youth identify as transgender in the United States?
Hutchinson B, Shircliff JE, Glass C, Miller GH, Marquez-Velarde G, Suárez MI (2024) Analyzing trans and nonbinary workers' response to workplace discrimination. Work Occup. https://doi.org/10.1177/07308884241240079
Johnson RG III, Renderos H, Gessese S, Kuboyama A, Liu E (2021) What is the ethical impact of gender non-conforming names? Pub Integr 23(6):541–554
Jones SE (2020) Negotiating transgender identity at work: a movement to theorize a transgender standpoint epistemology. Manag Commun Q 34(2):251–278
Kleintop LK (2021) Deciding to be authentic: transgender employees and their decision to be out at work. In: Exploring gender at work: multiple perspectives, pp 139–158
Köllen T, Rumens N (2022) Challenging cisnormativity, gender binarism and sex binarism in management research: foregrounding the workplace experiences of trans* and intersex people. Gend Manag Int J 37(6):701–715
Ladwig RC (2023) Managerial influences on the inclusion of transgender and gender-diverse employees: a critical multi-method study. Aust J Manag 48(4):693–710
Lambert JR (2015) The impact of gay-friendly recruitment statements and due process employment on a firm's attractiveness as an employer. Equal Divers Incl Int J 34(6):510–526
Majid A, Yasir M, Khan M (2023) Making decent work a reality to transgender individuals: a step towards a realization of sustainable development goals in developing countries. Pak J Commer Soc Sci 17(4):826–848
Mann S (2021) Transgender employment and gender marker laws. Labour Econ 73:102072
Mor K, Gündemir S, van der Toorn J (2025) Celebrating the "invisible": the role of organizational diversity approaches on attracting and retaining LGBTQ+ talent. J Bus Psychol 40(3):593–617
McFadden C (2020a) Hiring discrimination against transgender job applicants—considerations when designing a study. Int J Manpow 41(6):731–752
McFadden C (2020b) Discrimination against transgender employees and jobseekers. In: Handbook of labor, human resources and population economics, pp 1–14
Paniza MDR, Moresco MC (2022) On the margins of diversity management? Travestis, transsexuals and the world of work. Revista de Administração de Empresas 62:e2021–e0031
Roberson QM, Collins CJ, Oreg S (2005) The effects of recruitment message specificity on applicant attraction to organizations. J Bus Psychol 19:319–339
Rynes SL, Barber AE (1990) Applicant attraction strategies: an organizational perspective. Acad Manag Rev 15(2):286–310
Riha J, Ravindran TS, Atiim GA, Khanna R, Remme M (2025) Advancing gender equality in global health: what can we learn from successful gender integration across five UN agencies? PLOS Global Public Health 5(6):e0004697
Sabharwal M, Levine H, D'Agostino M, Nguyen T (2019) Inclusive work practices: turnover intentions among LGBT employees of the U.S. federal government. Am Rev Publ Admin 49(4):482–494
Salter NP, Sasso T (2021) The positive experiences associated with coming out at work. Equal Divers Incl Int J 41(2):224–240
Sawant NS (2017) Transgender: status in India. Ann Indian Psychiatry 1(2):59–61
Spence M (1978) Job market signaling. In: Uncertainty in economics. Academic Press, pp 281–306
Thomas KM, Wise PG (1999) Organizational attractiveness and individual differences: are diverse applicants attracted by different factors? J Bus Psychol 13:375–390
Thoroughgood CN, Sawyer K, Webster JR (2020) Creating a trans-inclusive workplace: how to make transgender employees feel valued at work. Harv Bus Rev

Thorsteinson TJ, Palmer EM, Wulff C, Anderson A (2004) Too good to be true? Using realism to enhance applicant attraction. J Bus Psychol 19:125–137

Walker HJ, Feild HS, Giles WF, Bernerth JB, Jones-Farmer LA (2007) An assessment of attraction toward affirmative action organizations: investigating the role of individual differences. J Organ Behav 28(4):485–507

Walia S, Mondal S (2025) Envisioning transgender equity: amplifying global south narratives through inclusive archives and education. J Int Women's Stud 27(3):1–41

Wille L, Derous E (2017) Getting the words right: when wording of job ads affects ethnic minorities' application decisions. Manag Commun Q 31(4):533–558

Wilson BD, Meyer IH (2021) Nonbinary LGBTQ adults in the United States

Zachariah M, Nandini AS (2023) Workplace transformation in India: creating a trans-inclusive environment. In: Managing for social justice: harnessing management theory and practice for collective good. Springer International Publishing, Cham, pp 309–343

Zimman L (2017) Transgender language reform. J Lang Discrim 1(1):84–105

Vibhor Kataria is Currently an Assistant Professor of Business Communication in HR and OB at FIIB, New Delhi, Vibhor is a corporate professional with 24 years of cross-functional experience across product development, project management, customer relationship management and sales. Having worked with global leaders in publishing—McGraw Hill Education, Macmillan Education and Pearson India Education Services, he has 17 years of experience in the publishing industry. Vibhor is a software engineer with a master's in computer science from East Tennessee State University, USA. Vibhor is passionate about academics and research in the area of employability.

Tavleen Kaur is an Associate Professor in OB/HR Area at Fortune Institute of International Business (FIIB). Having a Doctorate degree with 8+ years of experience in teaching & research, her work has been published in journals indexed in ABDC-A, B, Scopus & Web of Science. These journals of repute include International Journal of Conflict Management, Current Issues in Tourism, Journal of Electronic Commerce in Organizations to name a few.

Net Zero, Full Strategy: Integrating Environmental Goals with Business Strategy as Growth Engines

10

Shikha Shokeen and Alisha Khattar

Abstract

Sustainability is not optional for businesses anymore, considering the growing constraints and challenges posed by climate change, environmental degradation, and stakeholder demands. In order to make environmental stewardship a driver of innovation, resilience, and competitive advantage, this chapter discusses how environmental goals can be effectively incorporated into business strategy. The chapter charts the evolution of business cases for sustainability, beginning with cost cutting and legal obligations and ending with increased investor appeal and brand value. It also provides helpful advice on incorporating sustainability into operations, product design, marketing, supply chain, and corporate governance through strategic models, namely the Triple Bottom Line, Shared Value Model, and the circular economy principles. Successful integration and the lessons learned are of actual business practices of powerhouses such as Patagonia, Hindustan Unilever Limited (HUL), and Tech Mahindra. This chapter also discusses practical issues such as balancing short-term profitability with long-term sustainability and the dangers of greenwashing. Ecological goals are now an element of company growth and have grown synonymous with it as the environment changes. It also presents responsible environmental integration as a basic necessity for companies which are value-driven and future-ready, apart from being legal requirements.

S. Shokeen (✉)
Asistant Professor, Department of Business Administration, Maharaja Surajmal Institute, New Delhi, India
e-mail: shikhashokeen@msijanakpuri.com

A. Khattar
Student, MBA (Marketing), Rukmini Devi Institute of Advanced Studies, New Delhi, India
e-mail: alishakhattar04@gmail.com

B. Kaur et al. (eds.), *Integrating Industry 5.0 and 4.0 for Technology and Humanity*, Contributions to Environmental Sciences & Innovative Business Technology, https://doi.org/10.1007/978-981-92-1780-9_10

Keywords

Greenwashing • Ecological responsibility • Carbon footprint • SDG—sustainable development goals • Management

10.1 Introduction

The globe is at a turning point in the twenty-first century, where environmental deterioration, scarcity of resources, and strict regulations are present issues that must be resolved immediately. Along with an increase in greenhouse gas emissions, extreme weather events becoming more frequent, rising sea levels, biodiversity loss, and a host of other related socioeconomic problems, the average world temperature is rising. At the same time, the globe is using our natural resources—forests, freshwater, and fossil fuels in an unsustainable way, which is placing a serious strain on ecosystems and economies as a whole.

Under the pretence of more stringent environmental laws and compliance, governments and organizations are tightening their grip on corporations, which is even more unsettling given the prevailing cynicism. According to this way of thought, sustainability is now required for both the earth and sustainable businesses, rather than being an option. For several decades, boardrooms have mainly ignored environmental and social considerations such as "nice to haves," thinly veiled PR campaigns, humanitarian endeavors, or CSR compliance, all of which, in general, have little teeth.

Environmental, social, and governance (ESG) metrics were viewed either as secondary to core financial achievement or as continuous priority issues. Instead of being viewed as an investment that could spur innovation and improve resilience, efforts to reduce carbon emissions, use renewable energy, or implement circular supply chains were perceived as imperiling growth. However, this way of thinking is slowly shifting. Profit and the environment are no longer at odds. This is a time where every company all around the world has started considering sustainability as the main part of business strategies, not only to look good in the eyes of different stakeholders, but to make business more profitable. They've seen how adopting eco-friendly practices can cut down costs through better energy use and open up new opportunities to create sustainable and innovative products. More importantly, they now realize that focusing on sustainability helps their business grow and stay strong in the long run.

As surprising as it may be, today's younger generations like Gen Z and Millennials are able to make a big change by simply speaking up about the issues that are impacting their lives and may affect the lives of others as well. This has led to companies being more open and taking more responsibilities through a new combined group of stakeholders, customers, investors, and employees who prefer and support the brands that they align with in terms of values and beliefs.

In today's time, in order to make any decision, the investors are focusing extensively and closely on the environmental, social, and governance standards of an organization. Consumers have also shifted to brands that focus on the well-being of the planet. Investors have also shifted to the businesses that focus on the planet and its people along with good customer service. Various issues like government fines, loss of brand reputation or supply chain issues may arise if sustainability is not taken into consideration. Sustainable practices are also being encouraged and enforced around the world through the Government.

There exists intense pressure within organizations to minimize harmful impact on the environment along with financial and legal risks. This may be due to international agreements like the Paris Agreement and the Montreal Protocol, and comprehensive obligation to attain net-zero emissions. Simultaneously, cutting costs, lowering risks, and boosting innovation can be attained with the help of sustainability and compliance. By focusing on environmental sustainability an organization will be able to strengthen its competitiveness and build long-term value. It also helps by giving the firms an edge in the market due to generation of fresh product and process ideas.

An organization that has a strong pursuit of sustainability policies will be able to benefit more by identifying monetary gains and a positive environmental outcome. Various researches done in this field show us how an organization can perform better in the stock market, generate higher returns on investment, and become more resilient during economic downturns with the use of rooted environmental, social and governance practices. An organization will get operational advantages like minimized energy usage, minimized waste, improved supply chain efficiency, and reduced carbon emissions by merely following sustainable practices. Over time, an organization will be able to not only reduce costs but also enhance their profits with these improvements. Sustainability is viewed as a crucial element for innovation, growth and long-term competitiveness and now, not seen as a limitation over business growth.

The implementation of sustainability strategies still remains a major challenge despite the widespread understanding. Truly embedding sustainability into the core strategy of a company requires more than just policy changes; it demands a complete shift in organizational rethinking. This includes revising governance frameworks, recreating performance metrics, and realigning company culture and values; this is to support long-term goals rather than just short-term financial gains.

This requires knowledge sharing among different departments, active leadership, and consistent communication across all levels of management. With the help of this chapter, we aim to explain to businesses the importance of including environmental sustainability in their daily operations and overall core strategy. It focuses on the major factors supporting this shift, the impact of integrating sustainable practices, and what challenges a business may encounter while trying to balance financial goals with environmental responsibilities.

10.2 Methodology

Qualitative methodology is followed to analyze the integration of environmental objectives into corporate strategy as a stimulus for innovation and growth. Articles, industry reports, business sustainability reports, academic journals, and globally recognized frameworks were collected from various secondary sources. These resources were taken into consideration as they help by providing a thorough understanding of practical applications and theoretical frameworks.

Case studies have been taken from well-known companies like HUL and Tech Mahindra in order to demonstrate real-world applications. Thematic analysis technique is followed to analyze the content and to identify the reoccurring trends in sustainable operations. For the purpose of assessing alignment and impact, themes like supply chain sustainability, investor appeal, brand value, and a variety of other risks and concepts were mapped against strategic models. Reliability and applicability are ensured while getting the insights from multiple sources. Since this approach links academic concepts with real-world applications, it is suitable for evaluating how ecological responsibility is developing into a crucial element of corporate strategy that is future-ready.

10.3 Strategic Rationale

Integrating environmental sustainability into corporate plans is now a necessity rather than just a choice in the dynamic world of global business. Long-term sustainability, stakeholder dynamics, brand image, and financial performance are all impacted by environmental integration (Clark et al. 2015). This section discusses the important point related to different business cases for environmental integration, with emphasis on how they add significant value while promoting sustainability.

10.3.1 Risk Management: Climate, Regulatory, and Reputational

A crucial component of corporate risk management is incorporating environmental considerations. Businesses face two primary kinds of risks due to climate change: transitional risks and physical risks. Transitional risks are those risks which are related to changing laws, regulations, and consumer expectations. Physical risks are those risks which are brought on by extreme weather. Firms neglecting preparation for these changes risk financial losses and supply chain disruptions, which lead to reputational damages. A recent study analyzing social media data determined that reputational damage tied to ESG concerns can cause a substantial average drop of 0.29% in stock returns (Nicolas et al. 2023). This clearly demonstrates the financial repercussions of environmental failures. In today's digital era, where businesses and customers have the ability to hold corporations accountable, regulatory exposure seems to be a constant concern. Negative publicity about environmental negligence holds the power to undermine consumer trust and brand value. Moreover, businesses that openly disclose their environmental performance and exhibit responsible management cultivate a positive public image, hence protecting their market position in the long run.

10.3.2 Investor Pressure and the Rise of ESG Investing

According to Grand View Research (2024), the global ESG investing market was valued at $25.1 trillion in 2023 and is expected to grow at a compound annual growth rate (CAGR) of 18.8% through 2030 due to investors' growing emphasis on ESG factors. As demonstrated by organizations like the Church of England's Pension Board, which uses shareholder proposals to demand corporate ESG changes, this growth demonstrates that investors are active shapers of the ESG transformation rather than merely passive observers (Lumorus 2024). Additionally, a 2025 ProMarket study found that a 1% increase in sustainable debt investments results in significant organizational improvements, demonstrating the quantifiable relationship between the issuance of sustainable green bonds and other debt instruments and the improvement of corporate governance.

10.3.3 Brand Loyalty and Competitive Advantage

Patagonia's branding strategy emphasizes environmental concerns, and has educated consumers to invest more in ethically produced goods. Similarly, Tesla is boosted not just by its novel products, but also by its core commitment to accelerate the global move toward sustainable energy. An organization will get operational advantages like minimized energy usage, minimized waste, improved supply chain efficiency, and reduced carbon emissions by merely following sustainable practices. Over time, an organization will be able to not only reduce costs but also enhance its profits with these improvements. Sustainability is viewed as a crucial element for innovation, growth and long-term competitiveness and now, not seen as a limitation over business growth.

10.4 Case Analysis

10.4.1 Tech Mahindra's Sustainability Strategy

Tech Mahindra aims to be a trailblazer in sustainable digital transformation, by implementing sustainability into the very structure of its business across all levels, including operations, product development, and collaborations. It attempts at attaining net-zero emissions by 2040, accelerating innovation with green technology, and assisting businesses worldwide in achieving their own sustainability targets through low-carbon IT services. Their core strategy includes ESG principles, commitment to a resilient future by providing digital ecosystems, which are aligned with environmental regulatory guidelines. They have also committed to reducing their absolute greenhouse gas emissions from its operations by covering Scope 1, 2, and selected Scope 3 sources.

10.4.1.1 Integrating with Central Business Strategy

Tech Mahindra's main objective is integration of sustainability into its digital transformation services using ESG frameworks. They have provided its clients with various sustainability solutions, such as carbon footprint analysis, ESG data governance platforms, IT services.

Tech Mahindra has also integrated blockchain technology for auditable sustainable supply chains, as well as artificial intelligence (AI) for energy consumption optimization, into its green initiatives. As a result, Tech Mahindra has successfully redefined the concept of sustainable business growth by embracing digital innovation while also strengthening the company's competitive position as a global digital partner in the low-carbon supply field.

10.4.1.2 Internal Operational Sustainability Plan

Tech Mahindra aims at achieving sustainability through this plan, which promises to reduce carbon footprints while boosting its productivity for longer, leading to a positive impact, and aims at accomplishing this within the workplace, processes, and the culture of people. One of the important components of this is building green infrastructure and that includes its campuses such as Hyderabad campus which is a platinum-rated green building. The campuses are designed in a manner where the centers come with solar rooftop systems, energy-efficient lighting and HVAC systems and EV charging points which result in massive energy and greenhouse gas savings. Another priority is water conservation which is implemented in various ways including rainwater harvesting systems, and sewage treatment plants and gray-water recycling units have been established to warrant zero water discharge for all facilities.

Moreover, ethical disposal of e-waste via refurbishing, recycling, and donation of IT assets and improvement of waste segregation processes are also implemented to become landfill waste free.

10.4.2 HUL's Corporate Sustainability Strategies

Hindustan Unilever Limited (HUL) poses sustainable development as a vital component of its business model. The firm has recently developed a strategic framework, called Unilever Compass, which focuses on sustainable growth, climate action, and inclusivity. All operations of HUL's corporate levels are driven by their broader sustainability vision of making sustainable living commonplace.

10.4.2.1 Integration of Environmental Responsibility into Business Strategy

HUL has implemented sustainability as one of its key business strategies. It utilizes this to enable innovation which drives brand growth and long-term competitive advantage. This is because sustainability is no longer a mandatory legal compliance, rather it is an important element to incorporate into a firm's product design, marketing and supply chain decisions.

HUL aims at tackling broader social and environmental issues, which is usually evident in its household brands such as Lifebuoy, Surf Excel, and Domex. Lifebuoy raises awareness on hygiene and disinfection, while Surf Excel promotes water conservation through reduced rinse requirements, and Domex aids in spreading awareness on menstrual hygiene in underprivileged areas.

Apart from this, HUL also embodies sustainability in its operational side too, which can be seen in sourcing, manufacturing units, and is certified as zero waste to landfill, with over 60% of its agricultural raw material sourced sustainably.

10.4.2.2 Operational Sustainability and Green Manufacturing

HUL uses technology to strengthen its efficiency of operations and improve cleanliness, which is an extension of its mission to incorporate sustainability into its various operations such as logistics and manufacturing. This firm looks for ways in which it can adopt sustainability into its daily operations and a prime example of this would be it sourcing its majority of agricultural raw materials sustainably.

All this has helped HUL reduce waste, cut costs, and increase efficiency without sacrificing quality.

10.4.3 Patagonia's Environmental Friendly Brand Strategies

An outdoor clothing and equipment company founded in 1973, Patagonia is one of the most well-known examples of an environmentally responsible brand. Since its inception, this brand has led with the belief that business should not harm the environment. Patagonia has well-integrated sustainability into everything it does—from the raw materials to its marketing.

Its approach has inspired many other businesses to reconsider their growth and responsibility strategies. Patagonia is often seen as a global benchmark for sustainability and social responsibility in business. Its long-standing dedication to ethical manufacturing, using recycled fabrics, fair labor practices, and encouraging consumers' recycling and reuse of its own products instead of buying new ones, has put this brand in the folder of "Investor Favourite."

This brand has proven that it's possible to be profitable while staying true to strong ethical values.

10.4.3.1 Integrating Environmentalism into Business Strategy

Patagonia has been committed to environmental sustainability from the start of its business. They began the "1% for the Planet" project in 1985. Patagonia vowed to donate 1% of its total sales each year to the various environmental projects. Patagonia built on this social and environmental sustainability heritage when, in 2022, founder Yvon Chouinard donated the roughly $3 billion-dollar business by selling it to a trust and nonprofit dedicated to combating climate change, such that all profit not reinvested within the company will fund environmental causes and community assistance.

10.4.3.2 Pioneering Sustainable Marketing Campaigns

Patagonia's marketing efforts are founded upon encouraging sustainability and countering consumerism. The 2011 "Don't Buy This Jacket" campaign, where a full-page advertisement in one of the leading newspapers, challenged consumers to rethink unnecessary consumption and focused on product durability, encouraged consumers to avoid unnecessary purchase, and highlighted product longevity. This not only caused a controversy over overconsumption but also reinforced Patagonia's position in conservation.

In 2024, Patagonia released the satirical documentary "Shitthropocene," which points to the culture of over-fussing about clothing, including its own brand. The film is meant to stimulate behavioral change toward more sustainable behavior through the integration of expert information and humor.

10.4.3.3 Embracing Circular Economy Principles

Patagonia has encouraged the circular economy paradigm with programs such as Worn Wear, which asks customers to recycle, reuse, and repair. With repair programs and the sale of second-hand products, Patagonia extends the life of the garment, lessens waste, and diminishes the environment. While other aspects of its business model align toward a closed-loop model, Patagonia is committed to environmentally harmful fabrics, including recycled materials and organic cotton for its products. By using these materials, it reduces its supply chain environmental footprint and improves the industry's standards for sustainable material specification.

10.4.3.4 Transparency and Advocacy

Transparency is at the heart of Patagonia's organizational culture. The firm gives outsiders information about the supply chain, environmental impact, and areas where it needs to improve. Such transparency sets up a platform of trust among consumers and stakeholders and hence makes Patagonia a socially responsible company.

In addition to that, Patagonia has gained much accolade for participating in brand activism. Patagonia sued the United States government in 2017 over reduced national monument protections, thus indicating that it would fight for the protection of the environment through political activism.

10.4.3.5 Community Engagement and Education

Patagonia leverages its platform to educate and engage its community on environmental issues. Through storytelling, documentaries, and collaborations with environmental organizations, the company raises awareness and encourages collective action toward sustainability.

10.5 Challenges in Integrating Environmental Sustainability

10.5.1 Short-Term Financial Pressures Versus Long-Term Environmental Benefits

Organizations constantly face the challenge of reconciling short-term financial performance with the need to invest in long-term environmental sustainability initiatives. The need for short-term financial returns usually hinders the organization's ability to invest in sustainability initiatives that may not give results in the short-term, which is a challenge making it necessary to create a shift in the corporate strategy, resulting in appropriate valuation along with short-term financial performance.

As per EY, 78% of investors have a preference toward companies that focus on ESG activities, even if it hits short-term profits. According to PwC India research conducted with 180 C-Suite executives across manufacturing sectors, 93% of Indian manufacturers have sustainability targets with 2-3x profitable growth through Industry 5.0.

10.5.2 Lack of Knowledge or Training among Leadership and Staff

Effective implementation of environmental sustainability requires informed leaders and employees trained in sustainability. However, many organizations fall short in the knowledge and training required to support adoption of sustainable practices. The investment in education and training is essential to enable workers and leaders to make sustainable decisions.

10.5.3 Greenwashing: Superficial Claims without Real Commitment

Greenwashing, or the fabrication of false claims of environmental sustainability of a product or a firm, discredits real sustainability initiatives and can be regarded as the source of lost consumer trust. For instance, fashion companies have historically been the subject of criticism and legal action regarding exaggerated sustainability assertions. Regulatory authorities, including the European Union, have taken measures such as the Sustainable Finance Disclosure Regulation to counter greenwashing (Delmas and Burbano 2011).

10.5.4 Supply Chain Complexity

Current supply chains are complex, with diverse stakeholders in geographically dispersed areas, and sustainable practices, hence, become tough to execute. Environmental compliance throughout processes poses a challenge when there is little transparency and teamwork. IoT and Blockchain technologies are useful for tracing and authenticating sustainable practices throughout the processes.

10.6 Future Prospects for Innovation in Environmental Integration

10.6.1 Eco-Design and Sustainable Product Development

Sustainable design, or eco-design, brings environmental aspects together with product design to lower the negative impact at any point during a product's life. Such an approach comprises material selection, energy saving, recyclability, and general environmental performance. The UK has recently launched its first commercial-scale facility dedicated to post-consumer polyester recycling, alongside the establishment of the Circular Fashion Innovation Network (CFIN), aimed at fostering a national textile recycling infrastructure. These initiatives focus on processing substantial quantities of textile waste in response to the projected 92 million tons of textile waste produced annually within the UK.

10.6.2 Data Analytics for Environmental Performance

Implementation of environmental performance analytics facilitates the observation, examination, and optimization of companies' sustainability parameters. Smart analytics platforms, e.g., SAS Viya, provide facilities to deploy responsible AI and decision intelligence, and the analytics assist with processing vast datasets on environmental management. It is made possible for inefficiencies, environmental influence projections, and maximizing utilization of assets to be spotted and made effective.

10.6.3 Green Financing and Climate Risk Insurance

Green financing refers to allocation of capital to projects that yield benefits to the environment, such as renewable or pollution prevention. Green bonds, a prominent instrument in this domain, fund projects aligned with environmental objectives.

Fig. 10.1 Opportunities of innovation in environmental integration

Additionally, insurers are increasingly investing in low-carbon initiatives, with a BlackRock survey indicating that nearly all global insurers now have at least one low-carbon transition goal, a significant rise from 2% two years prior.

10.6.4 Digital Tools for Tracking Carbon Footprint

Technologies such as the Internet of Things (IoT) and blockchain are revamping the process of tracking carbon footprints. IoT devices make real-time data available through sensors, so users can see their energy consumption and greenhouse gas emissions, and make corrections as they monitor their impact. Blockchain technology supports the capabilities of transparency and traceability in supply chains to verify sustainable practices. OpenSC is a platform that allows consumers to verify sustainability claims of products in the supply chain by tracking and tracing products on the blockchain—to see where products came from, where they have been, and where they are going through a QR code scan that tracks their real-time location (Fig. 10.1).

10.7 Defining Environmental Goals in a Business Context

In today's business environment, environmental sustainability is no longer an optional pursuit but a strategic necessity. Organizations worldwide are under growing pressure from governments, investors, and consumers to demonstrate concrete commitments to environmental protection through the establishment of clear and measurable environmental goals (Green Business Benchmark 2023).

10.7.1 Understanding Environmental Goals

Environmental goals are precise outcomes that organizations aim to achieve to reduce their environmental footprint. These practices could be limiting GHG emissions, reducing waste, conserving water, or limiting the use of renewable energy (Machen and Associates 2023). For example, a company can have a goal to cut carbon emissions by 30% within five years, eliminate the use of single-use plastics, or be only sustainable by a certain year.

The establishment of environmental goals begins with a clear understanding of the organization's ecological footprint in its operations and throughout its supply chain. This is usually facilitated by Environmental Management Systems (EMS), which offer an organized approach to environmental risk identification, prioritization, and tracking of performance.

10.7.2 Frameworks and Standards Guiding Environmental Goal Setting

Various global standards and frameworks guide businesses in setting science-based, credible, and actionable environmental goals:

10.7.3 Principles for Effective Environmental Goals

Environmental goals must be aligned with the goals and reporting standards of the UN, and/or any other such body (Table 10.1). In addition to that, they should be:

Specific: An environmental goal should focus on a distinct environmental area. Instead of reducing emissions, it is better to specifically reduce Scope 1 and 2 GHG emissions by 2030 by 40% (Green Business Benchmark 2023);

Measurable: Set measurable targets. For example, it should be possible to reduce the consumption of water by 20% per unit of output, or recycle 90% of the waste from production;

Achievable: Of course, the goal should be ambitious. However, it should be clear that it is possible to achieve the goal with the available resources and the company's potential. Otherwise, it will be impossible to maintain the credibility of the company and its employees (Machen & Associates 2023);

Table 10.1 Environmental frameworks and standards

Standards	Goal-setting guidance
Science-based targets initiative (SBTi)	• Aids firms in aligning emissions reduction targets with novel climate science • Targets validated are considered rigorous, ensuring alignment with a 1.5 °C or well-below-2 °C warming trajectory
Net zero goals	• Net-zero standard defines criteria for credible net-zero commitments across scopes 1, 2, and 3 emissions
United Nations sustainable development goals (UN SDGs	• Serves as a global roadmap for sustainable development • SDG 12 ("responsible consumption and production") encourages businesses to minimize waste, use natural resources efficiently, and report on sustainability metrics
Environmental, social, and governance (ESG) reporting	• Provide a means for companies to disclose their environmental, social, and governance performance to investors and stakeholders • Reporting standards such as global reporting initiative (GRI) and sustainability accounting standards board (SASB) help ensure transparency and comparability across industries

Relevant: The goals should make sense for the industry in which the company operates and the main areas of the environmental impact. For example, for logistics companies, it may be reducing the emissions of the vehicle fleet, and for IT companies—the amount of energy used to run data centers;

Time-Bound: With deadlines, each of the goals should have a set of deadlocks or dates. This will make them more accountable and easier to evaluate (SBTi 2024a, b).

10.7.4 Integration into Business Operations

Environmental goals should form part of the central business strategy and not be an isolated sustainability project. This needs to be led by the very top leadership of a company and involve employees at all levels (Machen & Associates 2023). Firms that succeed in embedding sustainability in their procurement, operations, product design and marketing are more likely to succeed in their environmental goals.

Analytical techniques and tools are important for reporting and environmental performance management. New technologies such as the Internet of Things (IoT), artificial intelligence (AI), and blockchain allow companies to track resource usage in real-time and trace products in the supply chain, facilitating more accurate reporting and faster adjustments.

Organizations must periodically review their performance and undergo external audits in order to track environmental progress toward goals and ensure accuracy of reporting. A standard such as ISO 14001 stresses the need for continuous improvement and promotes regular evaluation of environmental performance.

10.8 Conclusion

Environmental sustainability has now moved to mainstream corporate strategy from being a voluntary interest earlier mooted at the margins of consideration, especially with the increase in pressure from society, customers and investors for greener business practices. In the present time, we can notice that there is a conversion of the business world from environment as an externality and administrative concern to a world focusing on value creation with sustainability in its core. Nowadays, we have to take into consideration various factors like climate change, regulatory command and control, investor expectations while calculating whether a business can be considered a good business. Being profitable and chasing societal benefits are now considered as two separate entities for a business in order to achieve sustainability goals and capitalism in a high-purpose organization.

With better sustainability practices in an organization comes various advantages like reduced expenses, increased goodwill and a higher investor confidence. With the globally acclaimed sustainability plan by Tech Mahindra, one can easily identify how implementing sustainability strategies can be beneficial for both the organization and the environment. We should view sustainability as our helping hand rather than viewing it as a restriction as it helps in creating a long-term value among stakeholders apart from progressing the environment.

International treaties like the Paris Accord and United Nations Sustainable Development Goals helped in the formation of frameworks like Triple Bottom Line and Environmental Social Governance in order to improve transparency and benchmarking. It can also be used to establish measurable environmental goals. Organization at times may prefer short-term profits, lack leadership commitments, have complex global supply chains, or practice greenwashing will lead to major challenges (Ghobbe and Nohekhan 2023). But with the help of advanced technology and cross-sectoral collaboration, an organization can increase transparency and accountability in environmental performance tracking along the supply chains through visionary leadership.

With the growing changes in today's time, all corporate executives must possess the required knowledge about environmental intelligence along with their business intelligence in order to perform efficiently. The new-age leaders and motivators will have to incorporate an eco-conscious culture in organizations which also includes decision-making and appraisal mechanisms. These leaders have to meet high expectations of meeting statutory sustainability goals as well as nurture this planet and its people by generating socially responsible innovation.

Table 10.2 Firms and their sustainability plans

Firm	Sustainability plans
Tech Mahindra	• Net-zero emissions by 2040 • Green tech innovation and low-carbon IT services • ESG integration with core business strategy • Board-level ESG governance and external audits (CDP, DJSI) • Green infrastructure with IGBC platinum-rated campuses • Solar systems, EV charging, zero water discharge • Waste-free operations and e-waste recycling • Employee sustainability culture through "green Army"
Hindustan Unilever limited (HUL)	• "Make sustainable living commonplace" vision via Unilever compass • Purpose-led brands addressing social/environmental issues • 60% sustainable agricultural sourcing • Zero waste to landfill manufacturing • Renewable energy adoption (solar, biomass) • Water-positive operations with recycling systems • Shakti/Prabhat programs empowering 2.5 M+ women entrepreneurs • Responsible sourcing and human rights compliance
Patagonia	• 1% for the planet initiative (1% of sales to environmental causes) • $3B company donated to climate change trust/nonprofit (2022) • Anti-consumerism marketing ("Don't buy this jacket") • Circular economy through worn Wear program • Recycled materials and organic cotton usage • Supply chain transparency and impact reporting • Environmental activism and legal advocacy • Community education through documentaries and storytelling

So, with the help of this paper we were able to reiterate that a business strategy without environmental sustainability will be chaotic; so an organization must incorporate it as it is an ethical necessity. Avoiding liability, open markets, drawing talent, and driving sustained growth will come easily to those who choose the right path of business strategy incorporated with sustainability. A business should put more time into developing activities that help in aligning profits with the well-being of the planet as the sooner an organization takes this up, the better and greener the future economy will be.

Table 10.2 summarizes how companies like Tech Mahindra, HUL and Patagonia have aligned their visions with environmental sustainability, and how that has manifested into efforts and targeted goals.

References

Clark GL, Feiner A, Viehs M (2015) From the stockholder to the stakeholder: how sustainability can drive financial outperformance, 5 March 2015. https://doi.org/10.2139/ssrn.2508281

Delmas M, Burbano VC (2011) The drivers of greenwashing. Calif Manag Rev 54(1):64–87

Ghobbe S, Nohekhan M (2023) Mental perception of quality: green marketing as a catalyst for brand quality enhancement. https://arxiv.org/abs/2312.15865

Grand View Research (2024) ESG investing market size, share & trends analysis report. Grand View Research. https://www.grandviewresearch.com/industry-analysis/esg-investing-market-report

Green Business Benchmark (2023) How to set effective ESG goals and targets. https://www.greenbusinessbenchmark.com/archive/esg-goals

Lumorus (2024) The ESG revolution: how investor pressure is driving corporate sustainability. https://www.lumorus.com/the-esg-revolution-how-investor-pressure-is-driving-corporate-sustainability/

Machen & Associates (2023) How to set effective ESG goals and targets. https://info.machen.cpa/blog/how-to-set-effective-esg-goals-and-targets

Nicolas M, Desroziers A, Caccioli F, Aste T (2023) ESG reputation risk matters: an event study based on social media data. Financ Res Lett, 59:104712. https://doi.org/10.1016/j.frl.2023.104712

Science-Based Targets Initiative (SBTi) (2024a) Net-zero standard. Retrieved from https://sciencebasedtargets.org/net-zero

Science-Based Targets Initiative (2024b) How it works. https://sciencebasedtargets.org/how-it-works

Shikha Shokeen is an Assistant Professor at Maharaja Surajmal Institute with over 11 years of experience in teaching and research. She is currently pursuing her Ph.D. in Management from Guru Gobind Singh Indraprastha University (GGSIPU). Her research interests are multidisciplinary focusing on HR, Organizational Behavior, Education, and Marketing.

Alisha Khattar is an MBA (Marketing) student at Guru Gobind Singh Indraprastha University with a dynamic background in entrepreneurship and marketing. She has successfully founded and managed two businesses in the jewellery and beauty sectors, blending creativity with strategic business acumen. She has also authored research papers focusing on business innovation and sustainability.

Innovating for a Sustainable Future: The Interplay of Green, Socially Responsible, and Sustainable HRM Practices

11

Ankita Bansal and Anjali Vyas

Abstract

Organizations, in their efforts to meet the sustainability goals set worldwide, have no option but to adopt the environment and society-friendly practices while at the same time getting pressure. Nevertheless, businesses are not fully unlocking their ability to innovate and perform through the integration of green HRM (GHRM), socially responsible HRM (SRHRM), and sustainable HRM (SHRM) that are very poorly connected and implemented at times. Moreover, one of the key motivations behind the present research is to study the overlapping areas and integration of these three variables and to uncover their potential in terms of sustainability and innovation. Utilizing a PRISMA method, this research has analyzed 179 research articles dating from 2008 to 2025 to discover the trends, theme clusters, and their contributions to sustainability. According to the results, one of the most efficient methods to foster innovation is through GHRM, and at the same time, SRHRM and SHRM are becoming very popular due to the no-brainer of their practices being close to GHRM. The implications of this research are significant for both theorists and practitioners who need to consider the areas of innovation, green creativity, sustainability, and digital transformation as main drivers. The findings will be very useful for the HR practitioners, organizational leaders, and policymakers as well as the researchers who are trying to come up with the integrated, innovation-based HRM strategies that cover environmental, social, and economic sustainability issues.

A. Bansal (✉) · A. Vyas
University School of Business, Chandigarh University, Mohali, Punjab, India
e-mail: ankitabansal358@gmail.com

A. Vyas
e-mail: anjali.e12890@cumail.in

B. Kaur et al. (eds.), *Integrating Industry 5.0 and 4.0 for Technology and Humanity*,
Contributions to Environmental Sciences & Innovative Business Technology,
https://doi.org/10.1007/978-981-92-1780-9_11

Keywords

Green HRM • Sustainable HRM • Socially responsible HRM • Innovation • Bibliometric analysis • Value creation

11.1 Introduction

Environmental deterioration, civic duty, global warming, social discrimination, morality, anti-capitalism protests, and the emergence of influential voices have been significant topics of discussion (Aggerholm et al. 2011; Jabbour and Santos 2008; Yusoff et al. 2015). Researchers indicate that overusing precious resources is a major contributing factor to these catastrophes (Meadows et al. 2004); to eliminate further environmental degradation that could arise from the over-exploitation of natural resources, organizational investors must develop a strategy for a sustainable future and nurture environmental prosperity (Christmann 2000; Khanna et al. 2002; Shrivastava 1995). Studies show that human activities often trigger various forms of natural disturbances. Identifying and improving the psychological behaviors that contribute to these disturbances can be a fundamental strategy for eliminating them (Ones and Dilchert 2012; Oskamp 1995; Pavitra Mishra 2017). Likewise, it may be beneficial to organizations implementing environmental management practices, as studies have found a positive relationship between green activities and consumer perception, brand image, profits, competitiveness, potential cost savings, and promotional opportunities (Christmann 2000; Mishra 2017; Shrivastava 1995). Scholars conceptualized the term sustainable development with increasing concern over ecology, emphasizing the overwhelming need for organizations to consider how their contributions to the environment and society are even more important than economic performance (Bombiak and Marciniuk-Kluska 2018; Malik et al. 2020).

In the present scenario characterized by globalization and rapid technological advancements, corporations are steadily being forced to adopt sustainable practices in their human resource management (HRM) systems. All over the world, companies are having to deal with hurdles such as the necessity to keep their workers content, the obligation to shrink their carbon footprints, and the need to ensure that they are innovating (Kramar 2014). The human resources department is proactive in uniting the organization with a sustainability strategy that is inclusive of environmental, social, and economic considerations and is also supportive of pinpointing the strategic leadership skills, capabilities, and competencies (Rimanoczy and Pearson 2010; Yusoff et al. 2015).

The HRM function has undergone a transformation to the strategic function level, where sustainability and innovation are the main objectives, thus ensuring organizations' lives within the environmental and social limits (Ehnert et al. 2014). The interplay and connection among innovation, sustainability, and HRM are strong, while the areas of GHRM, SHRM, and SRHRM are critical for realizing environmental sustainability, responsibility, and employee involvement (Dumont et al. 2017; Renwick et al. 2013). These newly emerging concepts seek to synchronize organizational practices with a more comprehensive sustainable development aspect that encompasses ethical considerations and the governance of the environment, keeping innovation as the pillar for value creation and competitive advantage (Akbari et al. 2021; Singh et al. 2020). GHRM is one of the activities that encourages the adoption of green policies, such as recruitment processes and training programs that are environmentally friendly. The emphasis of SRHRM is on equal treatment in labor practices and employee welfare, and the sustainability aspect is regarded by SHRM as a source of increased flexibility and adaptability in the long run. Therefore, this highlights the emerging strategic view of HRM as an important catalyst for organizational resilience and adaptability (Ehnert et al. 2016).

Besides, the progression of various innovations like artificial intelligence (AI), HR digitization, and data mining of large datasets has changed the way HRM systems work in companies greatly and has enabled the companies to promote green programs, enhance employee satisfaction, and innovate with data-informed insights (Akbari et al. 2021).

11.2 Research Objective

The convergence of high-tech and eco-friendly practices will give rise to a new generation of HRM strategies that will not only improve the bottom line but also prioritize social and environmental considerations. Thus, to encourage a comprehensive investigation, the purpose of this research is:

Research Objective—RO1 is to look into the past, present, and future of these variables (SRHRM, GHRM, and SHRM) with the help of a bibliometric analysis providing insight into the number of publications and citations of the different variables over time, as well as the most influential ones in the research area.

This study further delves into the thematic clusters, central themes, and emerging trends that lead to a complete understanding of the impact of these HRM paradigms on organizational innovation and value creation.

Research Objective—RO2 is primarily concerned with the detection of the thematic clusters and core themes throughout the literature and their connection with the concepts of innovation and sustainability, putting more emphasis on the latter.

Research Objective—RO3 evaluates the extent to which innovation contributes to the implementation of those HRM practices that are in line with both environmental and social goal sustainability.

11.3 Literature Review

11.3.1 Green Human Resource Management (GHRM)

The phrase "Green Human Resource Management" (GHRM) quickly became one of the most used terms in the business area, and its importance keeps rising. It does not only foster an understanding of environmental problems but also contributes to the company's economic and social well-being and that of its workers in a wider sense (Ahmad 2015). The separation of GHRM practices from the organization's day-to-day activities can result in an increase in productive output with an increasing ratio of the former to the latter (Mathapati 2013), and alongside that, there will be a decrease in costs, and levels of motivators and engaged workers will be at their highest. Environmental impacts will also be more favorable (Arulrajah et al. 2016), and that will lead to organizational success. Thus, the environmentally friendly efforts can last through being the source of support for sustainable objectives in the HRM process, which includes recruitment, education, compensation, training and development, and the overall growth of the firm's human resources (Ahmad 2015; Dutta 2012). According to Bombiak and Marciniuk-Kluska (2018), environmental awareness and resilience are the main results of green initiatives, thus GHRM and its practices are not just essential but also very intrinsic to sustainability, at the same time allowing for new ideas from the organizations (Sharma and Gupta 2010).

11.3.2 Socially Responsible Human Resource Management (SRHRM)

SRHRM aims to mold the attitudes and conduct of staff, thereby contributing to the betterment of the overall social responsibility of the company—the internal as well as the external one (Luu 2021; Vu 2022; Zhang et al. 2022). Bowen (1953) first coined the term "Social Responsibility" in his book "Social Responsibility of the Businessman." The idea of social responsibility has been elaborated a lot (Basu and Palazzo 2008; Carroll 1998; Porter and Kramer 2002; Shen and Zhu 2011). Under increasing pressure, organizations are forced to be more responsible while still making profits (Lombardi et al. 2020). Shen and Benson (2016) defined SRHRM as "The practical action of HRM divisions adopted by organizations in fulfilling the successful implementation of CSR." It inversely relates HR practices to social responsibility, putting a stronger focus on "employee well-being," "diversity," and "inclusion," and "ethical practices," thus increasing awareness of the company's influence on its workforce (Vu 2022) and making the workplace more ethical and sustainable (Kundu and Gahlawat 2015). SRHRM, furthermore, could create an innovative atmosphere amid employees by considering them as the main driving force of organization where they do feel appreciated, respected, and energized (Janssen and Van Yperen 2004; Zhao et al. 2024). An organization that embodies such a workplace culture will probably end up facing higher levels of employee engagement, trust, and cooperation which are all crucial factors in building innovation (Janssen and Van Yperen 2004).

11.3.3 Sustainable Human Resource Management (SHRM)

SHRM is the concept that combines environmental, economic, and social sustainability that is incorporated into various HR practices as the main strategy for long-term success and employee welfare (Ehnert et al. 2014). SHRM is characterized by its innovative approach. Unlike classical HRM which is mainly concerned with efficiency and performance, it promotes responsible management of the present and future generations' needs (Kramar 2014). To put it differently, this strategy is not only allowing the emergence of an atmosphere, in which employee development and innovation are the key factors leading to the acquiring of a long-term competitive advantage (Muñoz-Pascual and Galende 2020). Besides ecological sustainability, it also accounts for social and economic factors that are very important for sustainable HR practices (Ehnert et al. 2016).

The results of previous studies in HRM and sustainability suggest that there is a common view that businesses must be ecologically friendly and socially responsible. Green human resource management and development aims at the total reduction of the organization's ecological footprint by integrating environmental policies into human resource practices such as recruiting, training, and performance evaluations (Duggal and Jain 2020). Also, socially responsible human resource management promotes a proper alignment of the company's human resources practices with its corporate social responsibility initiatives, thus providing an incubating psychological atmosphere for communication to take place across the organization and, consequently, for innovation and value creation (Zhang 2023). However, the SRHRM studies have been neglected when compared to the GHRM and SHRM research, indicating that further investigation and study are called for. The interrelationship between them and their impacts on society and business is hardly comprehended, and the blending of these HR paradigms might produce synergies that render each of them capable of reaping the full benefits (He et al. 2021). The combination of GHRM, SHRM, and SRHRM not only fortifies organizational resilience but also sets the stage for sustainable competitive advantage through innovation. Therefore, the present research intends to contribute to the aforementioned gaps through a comprehensive bibliometric analysis of the literature.

11.4 Research Methodology

11.4.1 Data Extraction

This section describes the methodology and exclusion criteria for identifying and categorizing articles included in the bibliometric analysis; to identify and categorize articles, we have followed an approach of the PRISMA flow diagram developed by Moher et al. (2009). The PRISMA process involved examining detailed criteria used for obtaining eligibility of records, as presented in Fig. 11.1. The dataset for this article was extracted using Elsevier's bibliometric database, i.e., Scopus. A literature review that describes the research process undertaken in terms of sources following the approach given by (Kraus et al. 2020) and a bibliometric analysis based on the methodology by (Akbari et al. 2021). The database of Scopus was expressly chosen because it meets high standards and has comprehensive coverage. Over 68.7 million records in Scopus contain an abstract and 90.6 million documents from 26,591 active peer-reviewed journals. Further, Scopus covers so many business and management journals, and it's a very friendly tool for retrieving data (Scopus 2023). Bibliometric reviews use bibliographic data to perform a quantitative examination (Akbari et al. 2021) and identify potential areas for a new study on definable themes. An insightful bibliometric study has been meticulously conducted, adhering to established guidelines that ensure relevance of the study (Marzi et al. 2025).

11.4.2 Keyword Selection and Search Results

Figure 11.1 shows a systematic two-stage data extraction process for identifying and refining research records from Scopus. In the preliminary stage, data was obtained from Scopus through a comprehensive keyword search, combined with Boolean operators "AND" and "OR," to refine the search results. This search ensures a broader yet precise capture of relevant records, and 238 records were attained. The screening was based on the criteria of publication time (2008–2025), and no documents were rejected at this stage to highlight the research work of the last two decades.

The second stage of search results ensures the quality and relevance and involves a screening process based on the subject area focusing on the fields of "Business management, economics, accounting, decision science, econometrics, finance, social sciences, arts, and humanities" ($n = 195$). Publication type screening, considering articles, conference papers, book chapters, and reviews, led to the exclusion of 2 records ($n = 193$). Further, to maintain the high content quality, language and content screening were applied, which evaluate the clarity and relevance of content. After scrutinizing the content based on titles and abstracts, considering only the English language, 14 records ($n = 179$) were excluded. By combining the Boolean search strategy with multi-layered screening criteria, this rigorous process

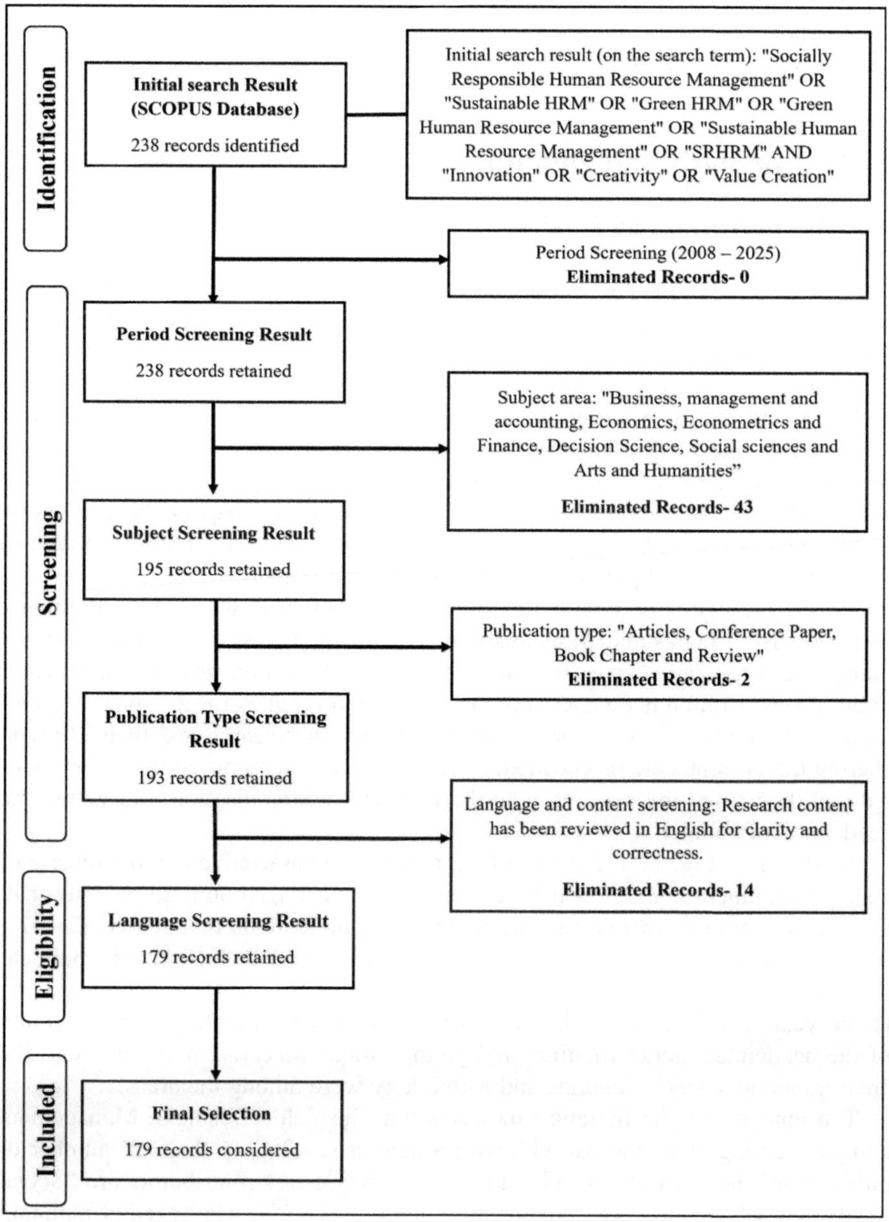

Fig. 11.1 Schematic representation of the article selection process. *Source* PRISMA 2020 flow diagram

ensured that only high-quality and relevant studies were included for further analysis, reflecting a thorough approach to data extraction and refinement. A total of 179 documents were ultimately considered for this study.

11.5 Analysis and Findings

11.5.1 Performance Analysis

The dataset covers the full period from 2008 to 2025 and thus provides a thorough view of academic performance through its metrics like specific disciplines, publishing patterns, and citations. A bibliometric literature review of 179 articles (see Table 11.1) pointed out that the relationship between innovation and GHRM, SRHRM, and SHRM has been under research since 2008 at the earliest. Each article, as a rule, has received an average of 33.69 citations. Moreover, the research is represented by 335 Keywords Plus (ID) and 509 author-selected keywords (DE). With contributions from 555 authors and 22 single-authored papers, the dataset displays a mixture of both individual and collaborative inputs. Regarding the type of documents, out of 179 publications, 146 are journal articles, which continue to be the primary source of peer-reviewed research that influences the field. Furthermore, there are also 4 books and 24 chapters included, which represent that the research output has been rather extensive and has appeared in more massive and longer publications. Conference papers account for 3, and reviews for 2, which together indicate the authors' awareness of the new trends in research and their effort in sharing the current state of knowledge. The document distribution shows how varied and diversified the research contributions and communications are within the academic community.

As shown in Fig. 11.2, the publishing trend was characterized by the release of a single document in 2008, which then remained unchanged until 2014. After that, there was a gradual increase in the number of publications that started in 2018, indicating a period of intensive expansion that lasted until 2020, when the publications suddenly rose to 66 in 2024, then continued with 1 in 2025, a good opening to the year. The factors that led to this development include higher participation of the academics, better funding, and giving global concerns more attention; the environmental sciences, energy, and technology were among the areas.

The analysis of the literature indicates that the field "Business, Management and Accounting" tops the list with over a quarter (24.2%) of the total number of publications, as illustrated in Fig. 11.3. The other major contributors are "Social Sciences" (17.4%) and "Environmental Sciences" (16.1%) who together maintain the importance of understanding social behaviors and dealing with the global problems of climate change and sustainability. The "Energy" theme (11.6%) highlights the growing global pursuit of green energy sources and technological advancements aimed at promoting sustainability. The domains' overlap represents the research field's complex connection and the common practice of sharing issues among fields to collectively find solutions to difficult, real-life situations.

Table 11.1 Main information about data

Description	Results
Timespan	2008:2025
Documents	179
Average citations per doc	33.69
Keywords plus (ID)	335
Author's keywords (DE)	509
Authors	555
Single-authored docs	22
Document types	
Article	146
Book	4
Book chapter	24
Conference paper	3
Review	2

Source Data from Scopus, processed using biblioshiny (R Studio)

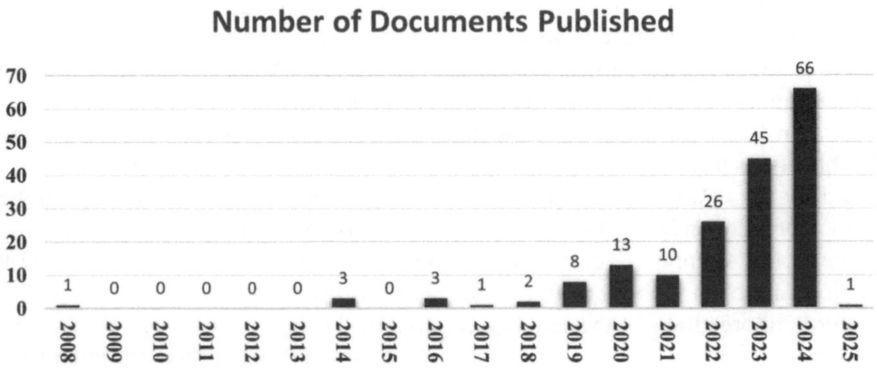

Fig. 11.2 Published documents. *Source* Scopus database

The research also offers the analysis of the leading authors, institutions, and countries; major articles, cluster analysis, co-occurrence, and thematic keyword analysis; and lastly, the identification of core themes based on keyword analysis in subsequent sections of the research.

11.5.2 Well-Known Authors, Institutions, and Countries

Table 11.2 analysis shows the most productive author is "Shafique Ur Rehman," with 4 publications and an impressive 522 citations. Shafique Ur Rehman stands out for his volume of work and also for the considerable academic influence; his

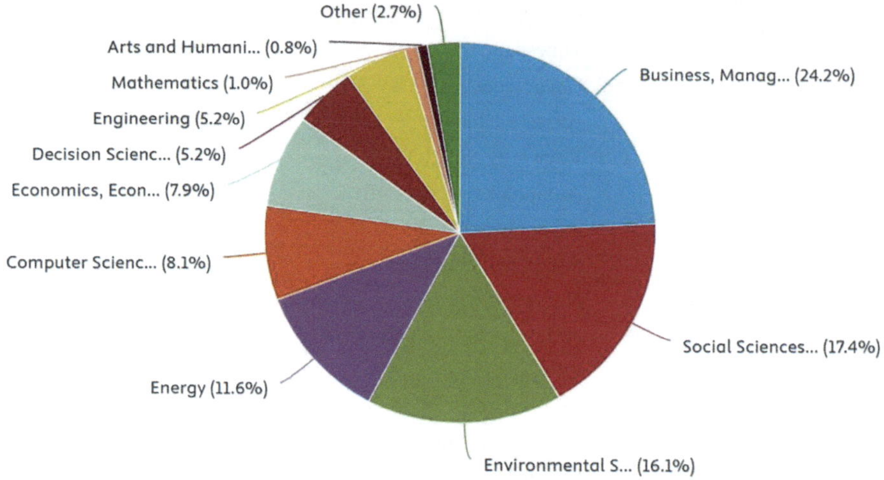

Fig. 11.3 Discipline-wise publication. *Source* Scopus database

study has a total citation count of 522, which demonstrates excellent academic acknowledgement, closely followed by Sascha Kraus with 2 publications and 499 citations, reflecting a significant presence. Paul Kivinda Muisyo, another prominent author with 3 publications, has received 267 citations.

In terms of institutions, the "College of Business at Abu Dhabi University," with 2 publications having 1185 citations, is the most influential institution. Closely followed by the "Paris School of Business," "National Research University Higher School of Economics," "University of Milano Bicocc," and "The University of Rome 'Link Campus'," each with 1 publication and citation scores of 1140 citations, reflecting a high degree of scholarly recognition.

Furthermore, Italy is the leading country in the list of countries with a total of 7 publications and 1899 citations, resulting in high-impact research output that is consistently done. China is the most productive country with 41 publications and 1460 citations. The USA, with 9 publications and 705 citations, demonstrates strong contributions too. The range of citation counts among different countries indicates the different levels of academic influence that each country has over the global research landscape. Pakistan, India, and Saudi Arabia are among the countries that are still significantly contributing to the field.

11.5.3 Most Influential Articles and Their Thematic-Based Clustering

The top 15 influential papers written by several authors are listed in Table 11.3. The article written by Singh et al. (2020) and having 1140 citations indicates the cooperation of green transformation and green innovation leadership, and Rehman

Table 11.2 Well-known authors, institutions, and countries

TC	Author	TP	TC	Institution	TP	TC	Country	TP
522	Rehman, Shafique Ur	4	1185	College of Business, Abu Dhabi University, United Arab Emirates	2	1899	Italy	7
499	Kraus, Sascha	2	1140	Paris School of Business, France	1	1460	China	41
267	Muisyo, Paul Kivinda	3	1140	National Research University Higher School of Economics, Moscow, Russian Federation	1	1265	Russian Federation	4
189	Qin, Su	2	1140	University of Milano, Bicocca, Italy	1	1261	United Arab Emirates	8
161	Song, Wenhao	2	1140	The University of Rome "Link Campus," Rome, Italy	1	1260	Pakistan	27
146	Ho, Thu Hau	2	452	Alfaisal University, Department of Management, Riyadh, Saudi Arabia	1	950	Saudi Arabia	23
146	Julius, Mercy Muthoni	2	452	Comsats University Islamabad, Business School, Attock Campus, Pakistan	1	705	United States	9
101	Soomro, Bahadur Ali	2	452	Free University of Bozen-Bolzano, Faculty of Economics and Management, Bolzano, Italy	1	509	Malaysia	25
77	Sobaih, Abu Elnasr E.	2	452	Ilma University, Ilma University Business School, Karachi, Pakistan	1	450	India	18
70	Piwowar-Sulej, Katarzyna	2	452	University of New Mexico, Anderson School of Management, Albuquerque, NM, United States	1	320	United Kingdom	16

Note(S) TC = total citations, TP = total publications
Source Authors' preparation from VosViewer output

et al. (2021) research of 452 citations is next to mention as it investigates GHRM's effect on environmental performance of manufacturing firms. Jia et al.'s (2018) and Farooq et al.'s (2022) research connects GHRM with leadership and promotes creativity and green viewpoint among workers; Ogbeibu et al.'s (2020) work on the other hand discusses the technological turbulence and its effects on sustainable HRM. Up next, Bhardwaj (2016) points out the SRHRM function in CSR and supply chain sustainability, whereas Joshi and Dhar (2020) present the positive impact of green training on capability development in traditional industries. To sum up, these studies signify the strategic role of HRM in attaining the organization's goal of sustainability and innovation through collaboration.

11.5.4 Thematic Clustering Through Bibliographic Coupling

Table 11.4 and Fig. 11.4 provide an overview of the three major clusters using bibliographic coupling, offering a structured perspective on the authors' contributions.

Green HRM, Innovation, and Environmental Performance: The collection of studies indicates the role of green human resource management in enhancing environmental performance and green innovation. HRM principles foster eco-friendly strategies that create value by promoting sustainable innovation. The study by Rehman et al. (2021) highlights the importance of organizations being active in participation in green innovation to successfully foster their green intellectual capital and GHRM, which ultimately improves their environmental performance and provides them with a competitive edge. An article authored by Bhardwaj (2016) discussed green policy together with CSR in sustainable supply chain management, emphasizing SRHRM's strategic role in embedding social responsibility in organizational processes. Further contributions by Aftab et al. (2023) and Song et al. (2021) discussed the mediating effects of green innovation and pro-environmental behavior, testing how these factors enhance environmental performance through proper environmental strategy and human willingness to take part in environmental activities. Thus, these publications emphasized the importance of innovation and how GHRM practices increase environmental performance in a quantifiable way in industries globally.

Socially Responsible HRM and Green Creativity: The articles that are the main focus of this cluster provide an excellent background for realizing the close relationship between SRHRM and innovation, especially in the areas of sustainable development and organizational performance. The studies done by Muisyo et al. (2022) and Muisyo and Qin (2021) are the ones that point out the moderating effects of leadership and organizational support on green creativity, thus highlighting the focus of sustainable HRM on employee engagement and innovation (Al-Hawari et al., 2021). Farooq et al. (2022) investigated the roles of GHRM and self-efficacy in producing green creativity in luxury hotels, and as a result, the service sector's distinctive difficulties were portrayed. Likewise, Ogbeibu et al. (2020) mentioned how technological turbulence influences team creativity and innovation in a short and comprehensive way, giving hints about the adaptability of SRHRM to the changing market conditions. Together, these publications constitute a powerful framework for understanding SRHRM, innovation, and green creativity interrelations.

Sustainable HRM, Leadership, and Organizational Performance: This theme focuses on the role of HRM in fostering green creativity and employee engagement

Table 11.3 Influential articles

Author(s)	Title	TC
Singh et al. (2020)	"Green innovation and environmental performance: The role of green transformational leadership and green human resource management"	1140
Rehman et al. (2021)	"Analyzing the relationship between green innovation and environmental performance in large manufacturing firms"	452
Jia et al. (2018)	"The continuous mediating effects of GHRM on employees' green passion via transformational leadership and green creativity"	212
Wikhamn (2019)	"Innovation, sustainable HRM and customer satisfaction"	185
Ogbeibu et al. (2020)	"Technological turbulence and greening of team creativity, product innovation, and human resource management: Implications for sustainability"	165
Aftab et al. (2023)	"Green human resource management and environmental performance: The role of green innovation and environmental strategy in a developing country"	158
Farooq et al. (2022)	"Do green human resource management and self-efficacy facilitate green creativity? A study of luxury hotels and resorts"	146
Song et al. (2021)	"Effects of green human resource management and managerial environmental concern on green innovation"	142
Munawar et al. (2022)	"Effects of green human resource management on green innovation through green human capital, environmental knowledge, and managerial environmental concern"	130
Muisyo and Qin (2021)	"Enhancing the FIRM'S green performance through green HRM: The moderating role of green innovation culture"	121
Stachová et al. (2019)	"External partnerships in employee education and development as the key to facing industry 4.0 challenges"	116
Waqas et al. (2021)	"Big data analytics as a roadmap towards green innovation, competitive advantage and environmental performance"	116
Bhardwaj (2016)	"Role of green policy on sustainable supply chain management: A model for implementing corporate social responsibility (CSR)"	116
Joshi and Dhar (2020)	"Green training in enhancing green creativity via green dynamic capabilities in the Indian handicraft sector: The moderating effect of resource commitment"	100
Hameed et al. (2022)	"How GHRM is related to green creativity? A moderated mediation model of green transformational leadership and green perceived organizational support"	90

Note TC = total citations
Source Author's preparation from VosViewer output

Table 11.4 Thematic clustering through bibliographic coupling

Cluster theme	Author	TC	Title
Green HRM, innovation, and environmental performance (red color)	Rehman et al. (2021)	452	"Analyzing the relationship between green innovation and environmental performance in large manufacturing firms"
	Aftab et al. (2023)	158	"Green human resource management and environmental performance: The role of green innovation and environmental strategy in a developing country"
	Song et al. (2021)	142	"Effects of green human resource management and managerial environmental concern on green innovation"
	Munawar et al. (2022)	130	"Effects of green human resource management on green innovation through green human capital, environmental knowledge, and managerial environmental concern"
	Bhardwaj (2016)	116	"Role of green policy on sustainable supply chain management: A model for implementing corporate social responsibility (CSR)"
Socially responsible HRM and green creativity (green color)	Ogbeibu et al. (2020)	165	"Technological turbulence and greening of team creativity, product innovation, and human resource management: Implications for sustainability"
	Farooq et al. (2022)	146	"Do green human resource management and self-efficacy facilitate green creativity? A study of luxury hotels and resorts"
	Muisyo and Qin (2021)	121	"Enhancing the FIRM'S green performance through green HRM: The moderating role of green innovation culture"

(continued)

Table 11.4 (continued)

Cluster theme	Author	TC	Title
	Joshi and Dhar (2020)	100	"Green training in enhancing green creativity via green dynamic capabilities in the Indian handicraft sector: The moderating effect of resource commitment"
	Muisyo et al. (2022)	78	"The role of green HRM in driving hotels' green creativity"
	Ahmad et al. (2022)	73	"The impact of green HRM on green creativity: Mediating role of pro-environmental behaviors and moderating role of ethical leadership style"
Sustainable HRM, leadership, and organizational performance (blue color)	Singh et al. (2020)	1140	"Green innovation and environmental performance: The role of green transformational leadership and green human resource management"
	Jia et al. (2018)	212	"The continuous mediating effects of GHRM on employees' green passion via transformational leadership and green creativity"
	Wikhamn (2019)	185	"Innovation, sustainable HRM and customer satisfaction"
	Hameed et al. (2022)	90	"How GHRM is related to green creativity? A moderated mediation model of green transformational leadership and green perceived organizational support"
	Sobaih et al. (2020)	77	"Influences of green human resources management on environmental performance in small lodging enterprises: The role of green innovation"

Note TC = total citations
Source Author's preparation from VosViewer output

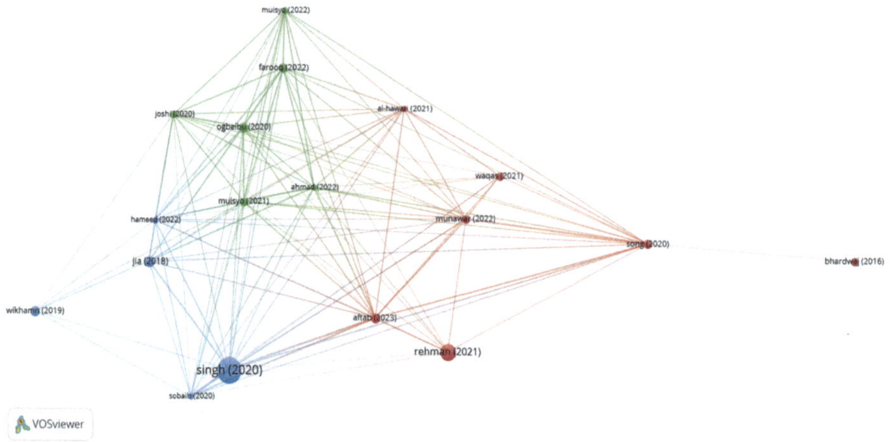

Fig. 11.4 Clustering via bibliographic coupling *Source* VosViewer output

with sustainability. These articles investigate the role of HRM practices in nurturing creativity and green innovation. Jia et al. (2018) highlighted the mediating role of GHRM in nurturing employees' green passion through transformational leadership and green creativity, highlighting how leadership influences creativity and sustainability. Sobaih et al. (2020) discussed the organizational culture and innovation has a role in moderating the effectiveness of GHRM. A study examined the role of transformational leadership and GHRM in achieving environmental performance and shows how strategic resources like leadership and HRM practices can be leveraged to enhance sustainability through green innovation Singh et al.'s (2020). A roadmap is presented through these publications for organizations to use SHRM and green creativity to achieve environmental sustainability and create an innovative and engaged working environment within the organization.

11.5.5 Co-occurrence Analysis of Keywords

Figure 11.5 is the co-occurrence analysis network of keywords focusing on articles with at least six occurrences. The prominent recurring keywords are "green human resource management," "green innovation," "green creativity," and "environmental performance," indicated by the larger nodes. From the co-occurrence analysis of keywords, we've established central themes among frequently used keywords in different key aspects of HRM. Four different clusters, each representing critical dimensions of GHRM, SHRM, SRHRM, and innovation were identified, as shown in Table 11.5. These clusters highlight the interconnections between HR practices, environmental sustainability, and innovation.

Cluster 1 revolves around the theme of "Digital Transformation and Sustainable HRM," which investigates the vital position of digital transformation in the

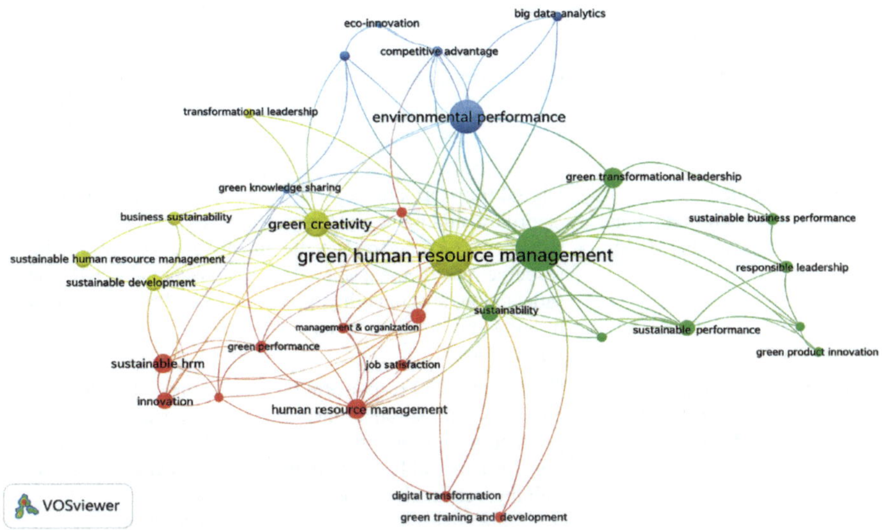

Fig. 11.5 Co-occurrence analysis *Source* VosViewer output

adoption of sustainable HRM practices. The cluster has associated keywords like "job satisfaction," "Digital transformation," "sustainability," "environmental management," "human resource management," "green behavior," "green training and development," "green performance," and "innovation"; therefore, it can be interpreted that organizations should open up to tech innovations not only for being eco-friendly but also for setting up more engaging employees all of which is in line with the earlier mentioned works that discuss the impact of digital transformation and environmental solutions (Hameed et al. 2022; Singh et al. 2020). Cluster 2 characterizes the subject of "Green Innovation and Responsible Leadership" with the focus on responsible leadership and CSR as the main factors which enhance green innovation. This cluster also contains the keywords such as "Corporate social responsibility," "green product innovation," "environmental sustainability," "sustainable performance," "green product innovation," "green training," and "green transformational leadership." The central topic indicates that SRHRM is a key factor in connecting employee development, engagement, and their well-being with sustainability. The theme supports some researches about HR practices affected by leadership-driven green innovation to sustainability (Awan et al. 2023; Dong and Zhong 2021; Ogbeibu et al. 2020; Singh et al. 2020). Cluster 3, the principal theme of "Big Data and Eco-Innovation," points to the technological aspect of environmental performance. Among the keywords in this cluster are "Big data analytics," "environmental performance," "competitive advantage," "green knowledge sharing," and "eco-innovation," which denote that the technology based on data is the protagonist for the advancement of eco-innovation and environmental performance (El-Kassar and Singh 2019; Waqas et al. 2021). Finally, Cluster 4 "Green HRM and Business Sustainability" as the central topic emphasizes the HR

Table 11.5 Identification of central topic and aspect of GHRM, SHRM, SRHRM, and innovation via co-occurrence of keywords

Basis	Central topic	Co-occurrence keywords	Aspects of green HRM, sustainable HRM, SRHRM, and innovation
Cluster 1	Digital transformation and sustainable HRM	Digital transformation, environmental management, green behavior, green performance, human resource management, job satisfaction, sustainability, green training and development, innovation	Integration of digital technologies in HR practices, promoting sustainable HRM and employee satisfaction
Cluster 2	Green innovation and responsible leadership	Corporate social responsibility, environmental sustainability, green product innovation, green transformational leadership, green training, responsible leadership	Leadership-driven innovation fostering SRHRM through ethical HR practices and sustainable leadership
Cluster 3	Big data and eco-innovation	Big data analytics, competitive advantage, eco-innovation, environmental performance, green knowledge sharing	Data-driven strategies enhancing green HRM, eco-innovation, and employee knowledge sharing
Cluster 4	Green HRM and business sustainability	Business sustainability, green creativity, green HRM, sustainable development, transformational leadership	HR-driven initiatives fostering green creativity, sustainable HRM, and responsible business practices

practices which are responsible for the eventuality of sustainability and innovation, this cluster has the keywords "Business sustainability," "sustainable development," "green creativity," and "green HRM" pointing that the companies are fine tuning their HR strategies to bring about the creative solutions for the environmental issues (Jia et al. 2018; Muisyo et al. 2022).

These clusters indicate how the integration of technology, leadership-based innovation, and creative HR practices promotes organizational sustainability and fosters an eco-culture.

11.6 Conclusion

This study provides a comprehensive analysis of the evolution, trends, and contributions of GHRM, SRHRM, and SHRM from 2008 to 2025 through a bibliometric study. The findings indicate a significant increase in interest.

The first research objective—RO1—investigates the progress and benefits of SRHRM, GHRM, and SHRM. This research finds that companies have acknowledged the role of HRM as a driver of sustainability in their operations and thus they have already very positive outcomes in relation to this topic. The evaluation showed that GHRM was inherently connected to the corporate environmental responsibility measures, whereas SRHRM has been related to the employee's well-being, ethical labor practices, and diversity and inclusion programs. Besides, SHRM has been associated with the development of a strategy for integrating sustainability into the medium and long-term HR planning and decision-making processes.

The second research objective—RO2—uncovers thematic clusters and core themes, particularly the connections between HRM, innovation, and sustainability. The findings highlight HRM's role in driving sustainability-enabled innovation through employee's active participation in green initiatives. The study further suggests that organizations implementing sustainable HRM practices tend to have a highly motivated workforce, a good corporate image, and more efficient operations. Besides the discussion about the overlap of the research field of sustainable HRM with environmental science, business strategy, and technology management, it raises a significant demand for interdisciplinary cooperation. These revelations call for a combined HRM model for the economic, social, and environmental sustainability triad.

The third and final objective of this study—RO3—is about the intervention of innovation and the way it supports HRM practices to maintain the balance between environment and society. The HR professionals in the companies that have introduced such technologies are opening up the doorways to a more professional and highly skilled workforce, which in turn, enables them to offer better HR services and outcomes. Digital technologies are applied to minimize the load of sustainability reporting, maximize the exploitation of resources, and create an eco-friendly workplace. Innovation is the driving force of ethical HRM practices that provide the well-being of the employees, the introduction of diversity programs, and the application of ethical recruitment practices. The results of the study imply that innovation is the source of greater productivity and a significant facilitator of environmentally friendly practices and socially responsible HRM.

11.7 Implications

The findings of this study have several significant implications for academic and organizational practice. RO1 investigates the evolution, trends, and contributions of SRHRM, GHRM, and SHRM; the study concludes that the intertwining of sustainable HRM practices with corporate social responsibility, environmental management; and digital transformation is gaining momentum. Organizations can not only align their HRM strategies to sustainability goals but also align their workforce management with sustainability, whereby scholars can further build theoretical concepts for sustainable HRM and their intersections with innovation and business ethics. RO2 reveals the role of HRM in promoting innovation and sustainability. The findings indicate that sustainable HRM is closely associated with employee engagement, leadership, and organizational performance. The study further suggests that future researchers may explore strategies for sustainable talent development, the adoption of green HR practices, and the creation of organizational cultures that foster innovation and encourage sustainable HRM, ultimately promoting a culture of environmental responsibility. Thirdly, RO3 assesses the impact of innovation on HRM practices related to environmental and societal goals. This research illustrates that firms must take advantage of the technology to ensure smooth and efficient HRM processes, along with keeping powerful environmental and social accountability. Interdisciplinary inquiries have been conducted on HR innovation's role in sustainability transitions.

11.8 Future Research

In spite of the thorough investigation, there are certain territories that are still unexplored in this regard. To start with, although the present-day literature mainly discusses the importance of green innovation and worker participation, fewer studies have tackled the issue of long-range consequences. The future inquiry should give more weight to the advantages of actions taken at the organizational level and the well-being of the employees. This would also provide a clear understanding of how the usual green HR system affects the aspects of motivation, job satisfaction, and retention of employees. One more point that can be mentioned for future exploration is the combination of AI and big data analysis in sustainable HRM. With digital tools being more and more accessible, it becomes necessary to know through studies how HR strategies powered by AI enhance sustainability, innovation, and employee performance. Moreover, the role that culture plays in the adoption and effectiveness of GHRM and SHRM in different areas and sectors is yet another area of research that needs attention. Apart from that, the bibliometric analysis was the central focus of the research study; thus, future research is planned to emphasize empirical inquiries using qualitative, mixed methods, case studies, and field experiments to understand it thoroughly. Last but not least, the impending investigation will assess disruptions like the changes brought in by the new regulation, and the global movements aimed at promoting sustainability, for instance, the United Nations Sustainable Development Goals (SDGs) and their impact on HRM practices, which would create a broad-spectrum evolution from sustainability to the behavior of HRM.

11.9 Limitations

This research has acknowledged, yet it has hidden limitations along with its contributions. Firstly, it is based on the Scopus database alone, which, though widely accepted as comprehensive, still lacks providing the perspectives obtained from other databases like Web of Science or Google Scholar. The future work might include a comparative analysis involving many databases that would allow the researchers to have a wider view. Bibliometric methods can identify research trends and key themes very efficiently but do not offer qualitative analysis; this limitation could perhaps be addressed in the best way by adding some qualitative research methods, such as interviews or case studies to the study for better understanding. Thirdly, the study is limited to the period from 2008 to 2025, which restricts the historical scope of the analysis. Future research may include earlier studies to provide a more comprehensive historical perspective. Furthermore, as the analysis was limited to English-language publications, the findings may be subject to language bias. Future studies may consider multilingual analyses to capture a broader range of perspectives.

Data Availability Statement The data has been extracted from the Scopus database and is available from the corresponding author. Datasets can be shared upon request with researchers.

Funding Statement This research received no specific funding from agencies, commercial, or not-for-profit sectors.

Conflict of Interest The authors declare that there are no conflicts of interest regarding this research.

Ethical Considerations This article does not contain any studies with human or animal participants.

References

Aftab J, Abid N, Cucari N, Savastano M (2023) Green human resource management and environmental performance: the role of green innovation and environmental strategy in a developing country. Bus Strateg Environ 32(4):1782–1798

Aggerholm HK, Andersen SE, Thomsen C (2011) Conceptualising employer branding in sustainable organisations. Corp Commun 16(2):105–123

Ahmad I, Ullah K, Khan A (2022) The impact of green HRM on green creativity: mediating role of pro-environmental behaviors and moderating role of ethical leadership style. Int J Hum Resour Manag 33(19):3789–3821

Ahmad S (2015) Green human resource management: policies and practices. Cogent Bus Manag 2(1):1–13

Akbari M, Khodayari M, Khaleghi A, Danesh M, Padash H (2021) Technological innovation research in the last six decades: a bibliometric analysis. Eur J Innov Manag 24(5):1806–1831

Al-Hawari MA, Quratulain S, Melhem SB (2021) How and when frontline employees' environmental values influence their green creativity? Examining the role of perceived work meaningfulness and green HRM practices. J Clean Prod 310(Aug 2020):127598

Arulrajah AA, Opatha HHDNP, Nawaratne NNJ (2016) Green human resource management practices: a review. Sri Lankan J Hum Resour Manag 5(1):1

Awan FH, Dunnan L, Jamil K, Gul RF (2023) Stimulating environmental performance via green human resource management, green transformational leadership, and green innovation: a mediation-moderation model. Environ Sci Pollut Res 30(2):2958–2976

Basu K, Palazzo G (2008) Corporate social responsibility: a process model of Sensemaking. Acad Manag Rev 33(1):122–136

Bhardwaj BR (2016) Role of green policy on sustainable supply chain management. BIJ 23(2):456–468

Bombiak E, Marciniuk-Kluska A (2018) Green human resource management as a tool for the sustainable development of enterprises: polish young company experience. Sustain (Switzerland) 10(6)

Bowen HR (1953) Social responsibilities of the business. New York, Harper & Row

Carroll AB (1998) The four faces of corporate citizenship. Bus Soc Rev 100–101(1):1–7

Christmann P (2000) Effects of "best practices" of environmental management on cost advantage: the role of complementary assets. Acad Manag J 43(4):663–680

Dong W, Zhong L (2021) Responsible leadership fuels innovative behavior: the mediating roles of socially responsible human resource management and organizational pride. Front Psychol 12

Duggal G, Jain LK (2020) Green HRM-A way to greening the environment. Globus An International Journal of Management & IT 11(2):1

Dumont J, Shen J, Deng X (2017) Effects of green HRM practices on employee workplace green behavior: the role of psychological green climate and employee green values. Hum Resour Manag 56(4):613–627

Dutta S (2012) Greening people: a strategic dimension Sumanta Dutta*. ZENITH Int J Bus Econ Manag Res 2(2)

Ehnert I, Harry W, Zink KJ (2014) Sustainability and human resource management: developing sustainable business organizations. Springer International Publishing

Ehnert I, Parsa S, Roper I, Wagner M, Muller-Camen M (2016) Reporting on sustainability and HRM: a comparative study of sustainability reporting practices by the world's largest companies. Int J Hum Resour Manag 27(1):88–108

El-Kassar AN, Singh SK (2019) Green innovation and organizational performance: the influence of big data and the moderating role of management commitment and HR practices. Technol Forecast Soc Chang 144(Dec):483–498

Farooq R, Zhang Z, Talwar S, Dhir A (2022) Do green human resource management and self-efficacy facilitate green creativity? A study of luxury hotels and resorts. J Sustain Tour 30(4):824–845

Hameed Z, Naeem RM, Hassan M, Naeem M, Nazim M, Maqbool A (2022) How GHRM is related to green creativity? A moderated mediation model of green transformational leadership and green perceived organizational support. Int J Manpow 43(3):595–613

He J, Morrison AM, Zhang H (2021) Being sustainable: the three-way interactive effects of CSR, green human resource management, and responsible leadership on employee green behavior and task performance. Corp Soc Responsib Environ Manag 28(3):1043–1054

Jabbour CJC, Santos FCA (2008) The central role of human resource management in the search for sustainable organizations. Int J Hum Resour Manag 19(12):2133–2154

Janssen O, Van Yperen NW (2004) Employees' goal orientations, the quality of leader-member exchange, and the outcomes of job performance and job satisfaction. Acad Manag J 47(3):368–384

Jia J, Liu H, Chin T, Hu D (2018) The continuous mediating effects of GHRM on employees' green passion via transformational leadership and green creativity. Sustainability 10(9):3237

Joshi G, Dhar RL (2020) Green training in enhancing green creativity via green dynamic capabilities in the Indian handicraft sector: the moderating effect of resource commitment. J Clean Prod 267:121948

Khanna M, Rose W, Anton Q (2002) Board of Regents of the University of Wisconsin System. Corporate Environmental Management: regulatory and market-based incentives. Source Land Econ 78(4):539–558

Kramar R (2014) Beyond strategic human resource management: is sustainable human resource management the next approach? Int J Hum Resour Manag 25(8):1069–1089

Kraus S, Breier M, Dasí-Rodríguez S (2020) The art of crafting a systematic literature review in entrepreneurship research. Int Entrep Manag J 16(3):1023–1042

Kundu SC, Gahlawat N (2015) Socially responsible HR practices and employees' intention to quit: the mediating role of job satisfaction. Hum Resour Dev Int 18(4):1–20

Lombardi R, Manfredi S, Cuozzo B, Palmaccio M (2020) The profitable relationship among corporate social responsibility and human resource management: a new sustainable key factor. Corp Soc Responsib Environ Manag 27(6):2657–2667

Luu TT (2021) Socially responsible human resource practices and hospitality employee outcomes. Int J Contemp Hosp Manag 33(3):757–789

Malik SY, Cao Y, Mughal YH, Kundi GM, Mughal MH, Ramayah T (2020) Pathways towards sustainability in organizations: empirical evidence on the role of green human resource management practices and green intellectual capital. Sustain (Switzerland) 12(8):1–24

Marzi G, Balzano M, Caputo A, Pellegrini MM (2025) Guidelines for bibliometric-systematic literature reviews: 10 steps to combine analysis, synthesis and theory development. Int J Manag Rev 27(1):81–103

Mathapati CM (2013) Original article green HRM: a strategic facet. Tactful Manag Res J 2(2):1–6

Meadows D, Randers J, Meadows D (2004) Limits to growth: the 30-year update. Chelsea Green Publishing

Mishra P (2017) A framework for sustainable organizational development in an emerging economy. Int J Organ Anal 25(5):762–788

Moher D, Liberati A, Tetzlaff J, Altman DG (2009) Preferred reporting items for systematic reviews and meta-analyses: the PRISMA statement. PLoS Med 6(7):e1000097

Muisyo PK, Qin S (2021) Enhancing the FIRM'S green performance through green HRM: the moderating role of green innovation culture. J Clean Prod 289:125720

Muisyo PK, Su Q, Hashmi HBA, Ho TH, Julius MM (2022) The role of green HRM in driving hotels' green creativity. Int J Contemp Hosp Manag 34(4):1331–1352

Munawar S, Yousaf DHQ, Ahmed M, Rehman DS (2022) Effects of green human resource management on green innovation through green human capital, environmental knowledge, and managerial environmental concern. J Hosp Tour Manag 52:141–150

Muñoz-Pascual L, Galende J (2020) Sustainable human resource management and Organisational performance: an integrating theoretical framework for future research. Small Bus Int Rev 4(2):e281

Ogbeibu S, Emelifeonwu J, Senadjki A, Gaskin J, Kaivo-oja J (2020) Technological turbulence and greening of team creativity, product innovation, and human resource management: implications for sustainability. J Clean Prod 244:118703

Ones DS, Dilchert S (2012) Environmental sustainability at work: a call to action. Ind Organ Psychol 5(4):444–466

Oskamp S (1995) Resource conservation and recycling: behavior and policy. J Soc Issues 51(4):157–177

Porter ME, Kramer MR (2002) The competitive advantage of corporate philanthropy. Harv Bus Rev 80(12):56–68, 133

Rehman SU, Kraus S, Shah SA, Khanin D, Mahto RV (2021) Analyzing the relationship between green innovation and environmental performance in large manufacturing firms. Technol Forecast Soc Chang 163:120481

Renwick DWS, Redman T, Maguire S (2013) Green human resource management: a review and research agenda*. Int J Manag Rev 15(1):1–14

Rimanoczy I, Pearson T (2010) Role of HR in the new world of sustainability. Ind Commer Train 42(1):11–17

Scopus (2023) Scopus content coverage guide. https://assets.ctfassets.net/o78em1y1w4i4/EX1iy8VxBeQKf8aN2XzOp/c36f79db25484cb38a5972ad9a5472ec/Scopus_ContentCoverage_Guide_WEB.pdf

Sharma R, Gupta N (2010) Green HRM: an innovative approach to environmental sustainability

Shen J, Benson J (2016) When CSR is a social norm. J Manag 42(6):1723–1746

Shen J, Zhu CJ (2011) Effects of socially responsible human resource management on employee organizational commitment. Int J Hum Resour Manag 22(15):3020–3035

Shrivastava P (1995) The role of corporations in achieving ecological sustainability. Acad Manag Rev 20(4):936–960

Singh, S. K., Giudice, M. Del, Chierici, R., & Graziano, D. (2020). Green innovation and environmental performance: the role of green transformational leadership and green human resource management. Technol Forecast Soc Chang, 150, 119762

Sobaih AEE, Hasanein A, Elshaer I (2020) Influences of green human resources management on environmental performance in small lodging enterprises: the role of green innovation. Sustainability 12(24):10371

Song W, Yu H, Xu H (2021) Effects of green human resource management and managerial environmental concern on green innovation. Eur J Innov Manag 24(3):951–967

Stachová K, Papula J, Stacho Z, Kohnová L (2019) External partnerships in employee education and development as the key to facing industry 4.0 challenges. Sustainability 11(2):345

Vu T-V (2022) Perceived socially responsible HRM, employee organizational identification, and job performance: the moderating effect of perceived organizational response to a global crisis. Heliyon 8(11)

Waqas M, Honggang X, Ahmad N, Khan SAR, Iqbal M (2021) Big data analytics as a roadmap towards green innovation, competitive advantage and environmental performance. J Clean Prod 323:128998

Wikhamn W (2019) Innovation, sustainable HRM and customer satisfaction. Int J Hosp Manag 76:102–110

Yusoff YM, Othman NZ, Fernando Y, Amran A (2015) Conceptualization of green human resource management: an exploratory study from Malaysian-based multinational companies. Int J Bus Manag Econ Res 6(3):158–166

Zhang B (2023) Human resource management (HRM) practices and sustainable Organisational innovation: a staged exploration of the role of HRM based on a dynamic environment. Highlights Bus Econ Manag 3:157–169

Zhang Z, Wang J, Jia M (2022) Multilevel examination of how and when socially responsible human resource management improves the Well-being of employees. J Bus Ethics 176(1):55–71

Zhao X, Chutipattana N, Mohd-Shamsudin F (2024) Socially responsible human resource management and employee innovation performance: the roles of person-organization fit, work engagement and individualism orientation. Cogent Bus Manag 11(1)

Ankita Bansal is a seasoned professional with a strong academic foundation and diverse experience in Human Resources and Marketing. She earned her postgraduate degree in Business Administration with a focus on HR and Marketing from Chandigarh University and is currently pursuing her PhD there, specializing in HR research. Ankita is an Assistant Professor at Vivekananda Institute of Professional Studies and previously worked with Tech Mahindra as a Human Resource Business Associate, gaining valuable hands-on experience. Her commitment to social causes is reflected in her volunteer work with the NGO Manav Utthan Society (MUS India), where she dedicated her skills to impactful initiatives.

Anjali Vyas is from Jodhpur, Rajasthan, working as an Assistant Professor & MBA MOOC coordinator at Chandigarh University, Punjab. She started her career with consultancy company SVMARC in Delhi as an HR research analyst, where she worked on research projects linked with Wipro, Delhi Tourism, Revlon, etc. At a very young age, she achieved her doctorate in HR specialization. She is a distinguished scholar, being awarded as Young Researcher Award in 2023. Presently, she is supervising four research scholars at Chandigarh University. She possesses a collection of high-quality research papers in the areas of Green HRM, Organizational Citizenship Behavior, Employee Retention, Quality of Management Education, and Recruitment Analytics. One of her papers, which focuses on Generation Z employees, was awarded Best Paper at the International Conference.

GPSR Compliance

The European Union's (EU) General Product Safety Regulation (GPSR) is a set of rules that requires consumer products to be safe and our obligations to ensure this.

If you have any concerns about our products, you can contact us on ProductSafety@springernature.com

In case Publisher is established outside the EU, the EU authorized representative is:

Springer Nature Customer Service Center GmbH
Europaplatz 3
69115 Heidelberg, Germany

Batch number: 10732220

Printed by Printforce, the Netherlands